Toward Living Well with Less

Toward Living Well with Less
Roger Ulrich
Edited by Demetria Iazzetto

2012

Toward Living Well with Less
© 2012 by Roger Ulrich

Editor
Demetria Iazzetto

Cover and Layout Design
Claudia Lozano

Cover Photos
Claudia Lozano
Will Merydith
Jim

Proofreading
John Penn

Manuscript Preparation
Freddy Herrera

Gofer
Elizabeth de la Ossa

All rights reserved.

Behavior Development Corporation
7943 S 25th Street
Kalamazoo, MI 49048
www.lakevillagehomestead.org

First Printing: November 2012

ISBN: 978-0-615-69724-6

Printed in the United States of America

Articles in this compilation, plus additional articles on the subject, are online at: www.lakevillagehomestead.org/media

Contents

Foreword   ix
Preface   xi
Introduction   xv
  Demetria Iazzetto

PART I: STORIES FROM THE PAST

CHAPTER 1: Roots   3

  The C. M. Ulrich Family   5
  I Am Human Conflict and Adaptation   16
  Foreword from *Anabaptists in America*   47

CHAPTER 2: In and Out of Academia   51

  Memories from my College Years   53
  In Memory of Maybee Hall   62
  The Departmental Head   65
  Toward a More Perfect Union   81
  Animal Research: A Reflective Analysis   106
  Animal Research: A Psychological Ritual   112
  Animal Rights, Animal Wrongs and the Question of Balance   119

PART II: POINTS OF VIEW

CHAPTER 3: Education 133

   What Constitutes "Education"? 135
   Racism and Exploitation in Academe "Festering Out of Sight" 137
   Education Should Nourish and Promote Value that Exists in Farming 140
   Reevaluate What Being Educated Entails 143
   Just What Does "The Kalamazoo Promise"? 146
   Some Food for Thought 148
   Education Today Needs to Embrace, Foster Understanding of Agrarian Lifestyles 150
   Using Amish as a Guide, Not All Education Requires Classrooms and Degrees 152
   Universities Are Run Too Much like Businesses 154

CHAPTER 4: Lake Village Homestead 159

IN ROGER'S WORDS
   Interview with Roger Ulrich 162
   Beyond Walden Two: Living the Experiment 176
   Lake Village Homestead: A Letter to *The Michigan Land Trust Newsletter* 185

IN THE MEDIA
   Roger Ulrich and Lake Village Community: "The Experiment of Life" 190
   Field Trip Helps Cooper Students Learn about Agriculture 195
   Is It Utopia? 198
   Lake Village Homestead Lives On 206
   'This Is a Real Farm' 212
   Honesty, Respect among Lessons in Living 215

CHAPTER 5: Sustainable Living 219

   Treating Food as a Friend 221
   Human and Plant Ecology: Living Well with Less 225
   Urban Sprawl Fact: We Can't Eat Money 234
   On Preservation of Farmland and Open Space 235
   Can Congress Dictate What We Can Eat? 238
   Saving Our Countryside Holds the Key to Saving Our Cities 241

CHAPTER 6: Speaking Out 245

We Must Cut Racist, Exploitive Acts 247
Urban Sprawl and Its Effect on Farmland Continues Unabated 250
On Homelessness as Not Just a City Problem 252
Keeping Fowl in the City 255
No Need to Tear Down Waylee School 259
Time to Free Leonard Peltier 261
Eggs from Mentally Ill Chickens 263

CHAPTER 7: Guiding Philosophy 265

NARRATIVES

In Search of Our Achilles Heel 268
The Road to Cuba: An Essay on Some Predicaments of Modern Civilization 276
We Are All Related to One Another 290
What Goes Around Comes Around 292
The Use of Behavior Modification Strategies 293
Notes from a Radical Behaviorist 298

CREATIVE EXPRESSIONS

The Process Called Life 304
Genesis Revisited 305
Rites of Passage 306
A Gift from Gawath-ee-lass 308
Reflections from the Martyr's Mirror 309
The River of Relief 312

Foreword

There may be many ways to set the stage for revealing the essence of what Roger Ulrich's *Toward Living Well with Less* is about, but it is doubtful that any could top the contents of the following letter from Western Michigan University President Emeritus Diether Haenicke, wherein he open-heartedly concurs with Roger's guiding philosophy of environmental responsibility: "I have come over the years closer to your ideas that a return to a much simpler way of life with much more respect for our environment is needed."

April 30, 2008

Professor Roger Ulrich
Lake Village Homestead
7943 South 25th Street
Kalamazoo, Michigan 49048

Dear Roger,
    It is always a pleasure to get a letter from you, the last one with the many enclosures on life in the past, the Lake Village Homestead, and the C.M. Ulrich Family in particular.
    I read all of it with interest and regret that I never ventured out with my class when I taught courses about utopian societies. We always talked about the Oneida Community and other, largely religious, groups

who flourished in the Midwest in the 19th century, but I wish the students could have seen a real community that still attempts, and succeeds in, communal living. My failure!

I have come over the years closer to your ideas that a return to a much simpler way of life with much more respect for our environment is needed. I also see the enormous exploitation of our precious resources, our boundless greed for ever more, and the widespread senseless and appalling greed in our economic institutions and in our academic ones now, too.

But I also have become rather pessimistic about the likelihood of change in these matters. Most educators do not set an inspiring example for their students. Many of us are infected by the cancer of economic greed and social and environmental irresponsibility that I see everywhere. I am more or less resigned that things will take their course, that some sort of catastrophic shortages will eventually happen and will force us to rethink the fundamentals of our economic and environmental behavior. Nothing short of a real disaster will change our society, I fear.

Sorry to sound so glum, but best wishes and many blessings anyway, and thanks for writing.

Diether

Preface

When considering the here and now of life, it truly becomes important, in my opinion, to pay attention to the signposts that direct us toward the need to transition our daily actions in a way that enables our living well with less. After all the years of human activity here on Earth, it has become clear that the resources that maintain all life forms are not as abundant as was once the case. Indeed, more and more people are adding their voice daily to the call for all of us to begin to renew our efforts toward doing a better job at living more in balance with Mother Nature.

This book is a compilation of articles that tell the background story of my own transitioning that led to a return to the deep roots of my Amish Mennonite heritage, which resulted in the ongoing living experiment at the Lake Village Homestead farm community. Rolling Thunder summed things up in saying that one must live the truth to understand it. Words, after all, are just symbols whereby we communicate with one another our understanding of truth. All the stories in the following pages are just that: a point of view which attempts to, above all, be honest and provide a background of the truth of what led to and goes on at and in the Lake Village Homestead community.

What Lake Village is all about has been summed up in the words contained in the farm's brochure, "Agriculture that Supports the Community":

> The Lake Village Homestead Farm Cooperative grew out of an environmentally-oriented educational research program origi-

nally motivated by B. F. Skinner's book *Walden Two*. Presently this extended-family co-op occupies a mile and a half of lake shoreline and approximately 320 acres of forest, meadow, wetlands and farmland. The 160 acres of lakefront is owned by a nonprofit corporation dedicated to preserving the natural habitat and modeling sustainable agriculture. The farmland and shoreline are protected from non-farm development through enrollment in various local and state farm and open space protection programs.

The other acreage in the farm co-op belongs to individual families, many of whom at one time lived at the Lake Village Homestead and continue, in varying degrees, their association with its programs. Lake Village is an educational center for all ages that affords the opportunity for individuals to experience all aspects of building construction and manufacture, milking goats, feeding and watering livestock, grooming horses, collecting eggs, gardening, hunting, fishing, etc., while at the same time serving as a working farm.

By helping people gain a better understanding of their food source, Lake Village has served as a constant reminder that the sun, earth, air and water constitute our real wealth.

Lake Village is sustained on a day-to-day basis by individuals who live and work on the farm, while others live there and work elsewhere. Everyone, however, helps with chores and general homestead maintenance, and is dedicated to serving the land. Over the past thirty years, approximately 300 people have joined this intentional family-farm setting and contributed greatly to its survival. We consider ourselves fortunate members of the greater Kalamazoo community, which we respect as an integral part of nature's fabric. In this spirit, we work to promote the ecologically-sound environment necessary for our children's children to survive and reap the future blessing of a healthy Mother Earth. The land is carefully managed and shared among goats, chickens, ducks, turkeys, peacocks, horses, pigs, dogs, cats, cattle, people, non-domestic mammals, amphibians and reptiles. All animals are free to roam in their pastures. All gardening is done without the use of synthetic-chemical fertilizers, pesticides or herbicides.

For those interested in making a transition in their own lives toward living well with less, we submit that the earth, and the agriculture which it supports, in the final analysis nourishes our lives, and we stand ready at Lake Village to be of help to any and all who seek to make the move they know in their hearts needs to be made. If you feel we can be of mutual benefit to one another, get in touch.

With that being said, let me add a P.S. to this preface. Carla Emery, in her tenth edition of *The Encyclopedia of Country Living*, which has been called the original manual for living off the land and doing it yourself, lists the following suggestions:

> Find and buy your plot of land. Site and build a house or cabin. Create a vegetable garden. Log your land. Manage a herd of cattle. Raise chickens, goats and pigs. Mill your own flour. Bake your own bread. Sharpen an axe and all of your knives. Prune a tree. Harvest wood. Can, dry and preserve food. Drive a tractor. Treat your family with homemade remedies. Learn basic beekeeping. Light your table with candles you made.

The *Country Almanac* states that it is "packed with old wisdom as well as up-to-date websites and mail-order sources to make country living easier." *Organic Gardening* says, "If you're dreaming about moving 'back to the land' someday, or if you're already there and want to live more self-sufficiently [wherever you may be], you'll want… *The Encyclopedia of Country Living*. Carla Emery first started compiling *The Encyclopedia of Country Living* in 1974 and continued to improve, refine and update it through ten editions. She traveled the country, enthusiastically teaching the skills for living independently off the land—finding our place in the country, planning and planting a garden, raising animals, living the self-sufficient life. Whether you live in the city or the country—or anywhere in between—it is said that this book is the complete, practical guide to living well and living simply."

I would advise that the idea of living simply in practice is not simple. Rolling Thunder, a man native to this land, shared with us the indigenous idea that to understand the truth you must live it and warned that "it ain't easy." The transition to living well with less will require breaking many old habits that civilization has shaped us into believing must somehow be maintained, not as simple wants but as needs. Countless back-to-the-earth leaders have written countless books telling us over and over what we must do as we face the inconvenient truth outlined in Al Gore's book by that title. Here I would offer up as the best start the words of Tony Kaufman and design by Claudia Lozano in their booklet *The Simple Guide for staying fit, losing weight and improving relations with Mother Earth* and my own words in the foreword: Do It!

> Chapter One: Eat less.
>
> Chapter Two: Exercise more.
>
> Chapter Three: Choose healthy foods raised in healthy ways.
>
> Summary: For more information go outside and play, run, explore and then buy and eat some locally-raised healthy food.
>
> Author's Note: This guide, although brief, represents key findings from extensive research conducted by billions of people from all over the world for the past million or so years.

Introduction

It is my pleasure to introduce you to Roger Ulrich in these pages, in his own and others' words. I first met Roger on paper myself. He was one of several "Shapers at Work" profiled in the November 1972 issue of *Psychology Today*. The "shapers" were considered to be bright psychological thinkers associated with the work of behaviorist B. F. Skinner, expected to shape a better world through behaviorism.

My first introduction to B. F. Skinner was his novel *Walden Two*, which was a must-read during my undergraduate days in the early 1970s. Like many of my college peers, living in the turmoil of the times, I found Skinner's ideas about "living the experiment" an attractive, alternative view, promising a kind of earthly utopia, which suited my twenty-something-year-old idealism.

I would never have imagined that almost 40 years later I would have the opportunity to meet Roger, called one of Skinner's disciples—to become friends with him, to learn about his life and his work, and to be invited to recommend and edit the work presented in this collection. It has been fascinating work to do, and I trust that the decisions we have made together as to the book's content and organization do justice to Roger's life, work and vision.

I trust these chapters mirror well Roger's life journey, philosophy and lived practice. His awareness of the need for another big transition for people of the Earth is but a part of the fabric his words have woven here. In a letter to me dated June 2011, Roger wrote:

> Above all I want to emphasize that we are, I am, trying to live the truth of the need to transition toward living well with less.

I trust Roger's ideas will be seeds that you grow into something we can all be nourished by.

It is my pleasure to introduce you, through these pages, to Roger Ulrich.

Demetria Iazzetto, Editor
October 2012

Readers can find additional articles for Chapters 3, 4, 6 and 7 at: www.lakevillagehomestead.org/media

PART I

STORIES FROM THE PAST

CHAPTER 1
Roots

The themes in Roger's writings parallel his life history and form the structure for the chapters of the book, beginning with his Amish-Mennonite farm-oriented childhood in Chapter 1, "Roots." These articles document this period in Roger's early life journey. Invited to speak at a 1981 conference in Mexico on "Human Conflict and Adaptation in the City," Roger reflects upon the history of the Anabaptist movement in the U.S. and its dual commitment to nonviolence and nonresistance. He weaves together various experiences in his life journey, specifically his spiritual roots in the Amish-Mennonite tradition, his years of research as a behavioral scientist and his experiences as a member of the Lake Village Homestead community, along with reflections on the conference theme of human conflict and adaptation. To diminish human aggression, Roger recognizes, we need to begin by working on our individual selves. "Any change has to start at an individual level," he says. "We have to start teaching and living nonviolence at all levels."

◂ *With my grandfather, Christian H. Smith, in his backyard, just across the street from where I was still living after graduating from Eureka High School. His three acres were a virtual nature garden, containing flowers and all kinds of fruit and other trees, as well as a large truck patch in which he grew every vegetable imaginable and housed egg-laying hens.*

The C. M. Ulrich Family

Excerpted from The Illinois Mennonite Heritage Quarterly, *Vol. XXXIV, No.4, Winter 2007*

Part One: Grandpa C. M. Ulrich (1879-1945)

Grandpa C. M. Ulrich would come up the back steps to the bedroom where I slept with uncles Bob, Marvin and Jim, while banging on some pipes that ran to the bathroom below and shouting, "Boys, time to get up!"—and get up we did! No argument. Everyone just got up, quickly dressed and came downstairs to the breakfast that Grandma Martha had prepared.

    I don't remember much talk. We ate and then went out to do chores. Marv did the milking. Sometimes he would squirt milk right from the cow's tit into the open mouths of cats that stood in a circle around the cow. I also remember sometimes he would slap some cow that had offended him. I don't remember what chores Jim or Bob or Grandpa did specifically. Somehow eggs got collected, hogs got slopped, cattle and horses got fed and watered. Grandpa's team of horses was Maude, a white mare that I rode every chance I got, and a gelding named King.

    The barns were a haven of fun with tunnels in the hay and an old Klondike buggy in the machine shed, a relic of a day gone by before the advent of cars. All of this is a hazy, foggy memory of my early youth between birth and five. My aunts I don't remember in that context. Aunt Lit had left the home place, as did Mona, Lorene and Reva. As I grew older from 1938 on, more time was spent with Uncle Loren and Aunt Luella, testing the patience of these two saints as cousin Jack and I stretched the limits of freedom to roam and do as we pleased. We learned not to piss on the electric fence, how to hold young hogs that were having their nuts cut out, how to feed, kill and pluck chickens, how to drive tractors or a team of horses, and how to shock wheat and oats. We learned to be young members of a threshing crew on our way toward becoming useful members of the agri culture. Grandpa was always busy, not much for conversation or holding and hugging. That was Uncle "Tubby" Loren's role—always genuine, always fair and loving, even when he was trying to be stern as he kindly suggested that we behave differently.

Sometime in the late 30s or early 40s, Grandpa and Grandma moved to town, and Uncle Jim and Aunt Ruth took over the farm that had somehow been lost from family ownership to the bank and later purchased by Mrs. Sommers, a rich lady from Peoria who then hired the family to run it. So to Jim and Ruth fell the task of educating Jack and me on how to become farmers. My Dad, all the mechanics at the John Deere store, all the farmers who hung out there winter, spring, summer and fall buying tractors, tractor parts, pumps, getting tractors fixed, and just plain loafing, were my other early teachers. Once, when I was trying to be of help, I took a hammer and pounded the radiator of a John Deere. I was made to understand that there were other ways in which I could be more useful. I was absolutely aware that what I had done was not to be done again, while at the same time I can vaguely recall repressed smiles on faces that reflected love and understanding of a young never-to-be-mechanic. I was a horse man; Jack was the tractor guy. I mainly wanted to run foot races, play catch, ride horses and climb trees. The city water tower I left to Jack, who climbed the highest and fell the furthest—be it from the top of a silo or later in a snarled army parachute.

At the implement store, I once won a nickel for touching the electric contraption that the mechanics had built to zap flies that crossed its electrified bars to get to some dead meat in the bottom of the fly killer box, as I recall it. It was simply plugged into a wall socket and it hurt. I told my friend Johnny, who lived next door behind the barber shop, how the money-making scheme worked. He didn't seem to mind the shock as much and made many nickels, which I then won from him in the back alley on bets that I could throw a rock farther.

Dad and Grandpa talked a lot about deep issues I knew or cared little about—long conversations about how folks handled going bankrupt following the Depression. As I recall, Dad told me how John Colburn, the Eureka Bank president, had convinced Grandpa into mortgaging the home farm as the collateral for buying another 80 acres, which later could not be paid back, since farm produce was bringing in less than what it cost to produce. Long conversations about farmers not being able to get enough money from their crops to pay back bank loans necessary to keep the farms running were main topics of discussion during the 1930s, the era of John Steinbeck's book *The Grapes of Wrath*. So the tractors that Dad had sold couldn't be paid for because the farmers had little money. Later I came to understand this strategy of getting folks into debt is the same one that has been used by moneylenders and other economic hit men down through history. When you can't pay up, of course the folks in power suggest that you had better sell your farm, and if you refuse then the sheriff is called in.

So Grandpa lost the farm to the lady in Peoria, who Jim and Ruth now worked for, and Dad lost the John Deere business and went to work at Caterpillar, where he helped make tanks to help kill Japanese, Germans and Italians. Uncle Bob went west to California. I can still see him wildly driving up in front of the John Deere store shouting, "Slim, Slim, I killed Uncle Gus!" Uncle Gus Schertz, Grandma Liz's brother, had fallen off the combine Bob was pulling and the back wheel ran over his head. Bob cried, "Slim, I told him over and over to be careful." Farmers throughout history have been killed by machines and have taken their own lives after foreclosures, plus numerous other forms of agricultural genocide produced by our greed for money and our lack of understanding and respect for the earth, air, water and the people who use it ethically to feed us. The con job is the glorification of war as a means to get what we have been brainwashed by our educational institutions into thinking we must consume in order to be happy. Science, religion (which by the way are both faith-based ideologies although worshipped in slightly different churches), as well as our legal system, schools and corporations, all are a part of the same social setup of which we are all a part.

So Bob went to California, followed by Marvin, where they both trained to go kill the Japanese. In California, they found a place to live where they met Gladys Terp, who grew to love all Ulrichs. Marvin married a distant cousin, Maureen Sommers, who after becoming "great with a child" (Cousin Judy, who someone else had fathered) came out to California and married Marvin so as to legitimize her pregnancy. At that time, of course, we cousins knew little about all this because as young Amish Mennonites, we had not yet been introduced to what "worldly" really meant. (Looking closely at our current lives suggests that we would soon find out.)

Dad and Grandpa kept having their deep secret talks as to how to handle (I would guess) going broke, gossip about children marrying already pregnant women, and Uncle John Hamish, Grandpa's youngest sister Viola's husband, preaching sermons about "bad guys" joining the Army while the "good guys" became CO's [conscientious objectors] or got farm deferments. Dad frequently got so upset at Uncle John's sermons that he would simply bend over and lean against the pew in front of him and go to sleep—or probably, as I look back, pretend to. However, one time I do remember nudging him when he began to snore. Sometimes he would simply get up, walk out of church, go sit in his car and smoke his pipe. My Dad was the most honest, up-front person that has ever graced my life, and he is my all-time, number one hero. In short, he epitomizes fatherly love, which included the only spanking I ever got from him when I was about 2½. My Mom told him when he got home from work that I had climbed the fence and gone down by the railroad tracks. He picked me up by one arm, whacking my butt like I was a swinging target and booted me into the house through the screen door that a scared Momma Della held open. I got the message, and for the rest of my life I was careful to pay attention to his moods, which often signaled, "Go for it, Roger, even though there will be those who won't understand." He took me with him everywhere he could. We picked up eggs together driving for the Eureka Hatchery throughout several counties. He occasionally let me drive the truck and sometimes have a puff off his cigars or pipe. Years later when I told him about trying LSD, he just smiled and said, "What was it like?"

During my summer vacations, I traveled with him to every corner of the United States. We saw Sally Rand do her fan dance together at the Illinois State Fair in Springfield. I was nine or 10. "Now you don't have to tell your mother about this," he said with…a…grin on his face. He took me with him to meet Uncle Bob, along with Uncle Pete and Grandpa C.M., when Bob came home for a brief furlough before going overseas. Grandpa walked up and hugged him and started crying like I had no idea he was capable of doing. Through such observations made possible by being there, I was learning how these men loved each other deeply, without the verbal "I love you" which is more prevalent nowadays. My Dad, Slim, sometimes got very emotional. Surprisingly to me, he cried when our dog Bijou (Jewel in French) died after sister Carol backed over her with our car. I was awed by his sobs and remember him saying, "It's a damn shame it couldn't have been___," a distant relative whom he seriously disliked.

Dad loved to play ball, and I remember watching him playing first base on a city team when I was very little. He played catch with me a lot as did Uncle Pat, who was the best athlete of the uncles. Sports seemed to be everything to me, and between Pat and Dad I went to watch games everywhere. We were at Eureka College games as well as at Bradley from 1944 through 1949. I almost never missed high school football and baseball practices, managing to stand around in the field with the high school baseball players, prompting Coach Galbreth one day after I had made a good catch to say he wished they all took practice as seriously as did Roger.

During 1940-41, when Grandpa was dying with cancer, Dad would get me up late at night, and we would go over to his house on the north side of town and stand outside under Granddad's window and they would talk. Their bond was of a sort that I didn't understand at the time but have come to appreciate as I look back on mine with my Dad, that strengthens with time even after his passing.

My bond with my Grandpa C.H. Smith, with whom I spent much more time than with Grandpa C.M. as a function of living just across the street from Grandpa Smith, was truly special. Also, since he had only one grand-

David C. and Anna Reeser Ulrich family. Front: Viola, David, Anna, Emmanuel. Middle: Emma, Ella, Barbara, Anna, Elizabeth. Back: David, Christian M., Peter, Fanny.

son until cousin Wesley came along, Grandpa Smith and I had the opportunity for much more one-on-one interaction. It was his example and the love he taught me through it that I find myself replaying at age 74 with my own grandchildren, especially Rod and Kristina Gibbs' son, Alex. Grandpa Ulrich, on the other hand, had 12 grandchildren at the time he died. I also think Grandpa Ulrich's trauma at the death of Grandma Liz, and what he suddenly faced having to raise 11 children with the youngest being only about five, made the life he had less amenable to spending a lot of time playing. I felt that he and Grandma Martha truly loved one another, and in his final years he seemed to have more fun. Grandma Martha Ulrich was always very loving and caring of me. As she lay dying out in Virginia, she told me story after story of her life with Grandpa Chris and how thrilled she was when she found out that this handsome man had asked her boss at the retirement home if he thought she might be willing to marry him, knowing that it wouldn't be easy with all the kids he had. She smiled as she told me how, when she was asked, it was hard to not seem over eager and how much she loved Grandpa C.M. That too was some of what my Dad and Grandpa quietly discussed as they carefully maneuvered the ups and downs of family life. Father and firstborn son confided in each other things that could not be shared with others, certainly not publicly at church. When Grandma Martha died, Uncle Dick and Marvin put her in a coffin in the back of their station wagon and hauled her out to her funeral in Eureka. As I walked by the open casket, I took her hand in mine, kissed it and said, "See you later, Grandma." Aunt Ruth was right behind me… She said, "Roger, my goodness what are you doing…?" "Just saying goodbye to Grandma," was my reply.

One time, a Mennonite from Metamora bought a car from Dad. He traded in his old one and, when he drove it back to Heyl's to make the trade, Dad noticed that the tires that were there when the trade was made had been replaced with some old ones. As was often the case, I was hanging out around him at the garage at the time and, as I recall, the gist of his message to the Mennonite boy from Metamora was "to get…back home as fast as he could and change the tires back or the deal was off." He had a lot more to say after the young man left about having to be especially careful about trusting church people. To be honest, it was apparent to me that my Dad never really liked church and used every imaginable excuse he could think of to make Mom Della handle going to church with Sister Carol and me alone. To be honest, I didn't really like church all that well myself, although I was able to figure out ways to entertain myself that most of the time would fall short of the need for discipline of the type once meted out by Reverend Yordy, whereupon in the middle of his sermon he stopped and said, "Roger, you go sit over there, and Maurice (his often less-than-obedient son), go sit over there." We walked to opposite sides of the church as the congregation quietly watched and as our mothers prepared sermons that Maurice and I would later receive at home.

As I think back on my life with Dad, I can understand Uncle Jim's one time expressing his feelings of "poor Della" and "what she had to go through." Yet I know my Mom felt loved and lucky. Uncle Jim, like all my uncles, I loved like a brother. Mom would always say, when we would go to visit Grandpa and Grandma Ulrich, "Now you stay away from Jim because you are only going to end up crying." I, of course, ran to find Jim as soon as I got out of the car, and we went at it until I got an ear chewing or a Dutch rub or a roughing up that was over the top, and I, of course, cried—with, of course, no real sympathy from anyone. Roughhousing was simply a show of Ulrich love, and that love comes in a variety of ways. With Aunt Lit, it was much like that with Jim. We wrestled and, as we got older together, we talked and had fun, often staying and working for long periods of time at the various places she and Uncle Art farmed. I played ball with Uncle Art and Lit and broke my foot in a freak accident at Grandpa Garber's farm in Lowpoint. I traveled to fairs with them as they showed how their Belgian horses could drive with Uncle Art holding the reins. I remember the arrival of Phyllis, then Beth, then Jody, who I recall thinking were beautiful kids although not old enough to hang out with like I did with Aunt Lit. Probably of all my uncles and aunts, it was Jim, Ruth, Loren and Luella with whom I spent the most time. Indeed, Aunt Ruth and Aunt Luella were like additional moms, of whom I had many given that none of Mom's sisters had children of their own. Uncle Bob was the most like my Dad of all the brothers. Aunt Mona was sort of a tease that you never quite knew how to take. "What are you here for?" she might say if you came to visit. Aunt Lorene was more sober. Aunt Reva was full of fun and, like the whole clan, loved to tease and tell you how you should think, although none could match Uncle Dick in this category. The experiences I hold in my inner heart in regard to the tests that Uncle Richard put me through during all the days I spent under his and Aunt Elva's mentorships are legion and would fill a book. He was one of a kind. Once, Dad, Pat and Loren, so the story goes, pulled Dick up high into the loft of a barn over a big hole in the floor where the hay was dropped—it being the only way to the lower floor and the rest of the world.

The older brothers then let go of the rope onto which Dick clung high in the barn, thus releasing him to experience a 20-to-30-foot free fall to the bottom floor. Dick jumped up, grabbed a pitchfork and dared his older brothers to come down, swearing that he would gladly stab them in their collective behinds should they be fools enough to believe he was kidding. Dad told me that, knowing Dick, they were indeed believers and were held hostage there in the hayloft. As darkness came, Grandpa finally showed up as their savior, wondering what had happened that had caused his four oldest sons to miss supper.

Such was life on an Amish Mennonite farm sometime around about the Great Depression and beyond, and it is with great and respectful memories of the aunts, uncles, cousins, mothers, fathers, brothers and sisters that these words hopefully honor and pass onto our children's children the sense of caring I hold for them and for all our relations.

Part Two: Ralph "Slim" Ulrich (1904-1973)

I was walking toward Lake Michigan during the Chicago World's Fair in 1933 or 34 with my hand (so I thought) in my Dad's hand when I looked up and saw that the hand I was holding was attached to a total stranger who was smiling down at me. I turned and looked back and saw my Dad walking behind me with that big grin which was so much a part of him. I broke into a run back to him as fast as I could go back to my safety net. "My Father on Earth as He is in Heaven."

Dad's middle name, Crawford, came from Dr. Crawford, who home-delivered the Ulrich babies. Dad once told me that when Dr. Crawford came up the long drive in his horse-drawn buggy and would get out and go into the house to see their mother, they just knew that inside the black suitcase he carried was another baby brother or sister. Dad's nickname, "Slim," came from his always being thin when growing up. In his older years after picking up weight, people would ask why he was called "Slim." "It's because of my finances," he'd reply.

My home on Calendar Street was one block away from Dad's and Grandpa Smith's John Deere implement store. Mom was a stay-at-home mom. Slow and steady, "Dena" (Mom's nickname, which she never liked or allowed you to use) balanced an impatient, action-minded Slim, whose short fuse to explosion contrasted with my Mother's "never blow it" demeanor. Although we lived "in town" (Eureka at the time was about 1,300 population), our big fenced-in yard allowed for a pet dog (that was not liked by a neighbor who was suspect in her disappearance), chickens, a sheep for a while and rabbits.

The Methodist church next door was an outside-the-fence playground for me and a second cousin, Ronnie Ulrich, who lived just across the alley behind our home. He was seen by my Mother as an undisciplined troublemaker, which I suspect was simply a form of denial regarding my own role in the various pranks that got the two of us in trouble. Ron was afraid of Slim, whose teasing was more than Ron could handle, and thus often split the scene before Dad would come home from work.

At noon every day in Eureka, the fire sirens would signal the time as it also, with a different cadence, would signal a fire. Whenever that happened, my Dad and other Eureka Fire Department volunteers would rush from wherever they were toward the fire station, jumping up on the truck, putting on their waterproof uniforms and fire hats as they went. People would then follow to wherever the fire was and watch the firefight. Dad was also on the City Council and Chairman of the Lake Committee, which was responsible for establishing Eureka Lake as a new water source for the city. One night, Dad got a call from an angry lady who was a vehement opponent to the plan to build a lake for our drinking water. "Mr. Ulrich, I want you to know that I would rather drink the water from my toilet stool than from that new lake!" "Well, Mrs. Potts (can you believe the name, given this story?), all I can say is, wherever and whatever you are used to drinking is up to you," was Slim's gleeful reply, which got repeated over and over all about town. My Dad and uncles, as noted earlier, were roughnecks by nature but never mean—at least it never seemed that they meant any harm, even though at times, pain was the outcome. Uncle Bob and Reva's husband, Wilber, were play wrestling when, so the story goes, a leg was broken—whose I can't recall. When homemade ice cream was being churned, kids were encouraged to stick their finger in the hole leading to the ice from which the freezing excess water was meant to flow out from the wooden churn. If you could keep your finger there for a certain length of time, you could win a nickel and perhaps a frostbitten finger, should you "succeed" to meet the time frame.

Dad always smoked a pipe—had it going much of the time—always refilling and relighting it, often while talking and thus not noticing that the match had burnt down until its heat singed his fingers, causing him to jerk and throw the match. No smoking was allowed in our house, except in the basement, where Dad had his workbench and an easy chair near the coal bin from which we shoveled the coal to the furnace. It was in this basement where I practiced dribbling a basketball hour after hour from six years of age on up. Once I put a .22 bullet into the vise and then hit it, "Bang," with a hammer. Things like that made my Mom a bit nervous as to just what she could expect from her only son. In fair weather I played catch with Dad, who saw to it that the game often became one of burnout, which meant each of us would throw the ball at the other as hard as one could. I could tell early on that Dad was mindful of the fact that before too long he couldn't bum me out, which is to say that I could catch anything no matter how hard he threw it, unless of course it was a bit wild, which is what happened one day. He flung one past me into the side of our garage, breaking out a foot-long piece of siding that remained as a reminder of the event for years thereafter. Perhaps his wild pitch that day was what made him tolerant of the time Paul Leman threw a wild pitch that hit Dad, who was standing inadvertently off to the side, in his chin.

My Dad was most often supportive of what I wanted to do. His actions toward me exemplified the truth of the saying that it's not so much what we say to our children that is important, but rather how we make them feel. He made me feel safe. He got across a feeling of deep love that went far beyond any words he spoke. He had confidence in himself and passed the same spirit on to me. "Dad, what do you think I should do?" "I don't know. You're 21. Do what you feel is right." I wasn't actually 21. I was only about 13 when I first recall him giving me that line. Twenty-one came later.

My Mother's sisters, Mart, Lu and May, wanted me to go to Goshen College, from where Lu had graduated, and offered to pay my way. Dad knew I wanted to play intercollegiate sports, so he helped me get into North Central College, where I played four years of basketball and baseball, lettering three years in each. Often he would appear at games both at North Central and at other schools around the league. During summers, I played Sunday ball in a semi-pro league. Here again, he facilitated my participation in spite of or maybe because of the fact that some at our Mennonite church (mostly Mom's sister Lu) "frowned upon" playing ball on Sunday. "To heck with her," he would say. "They don't holler about May playing Sunday piano."

I'm 74 now and I am still playing ball. I stopped playing church a long time ago. I feel really good about myself—the life I have lived and am still living. I feel that is the spirit my Dad and Mom possessed and passed on to me.

Ralph "Slim" and Della Ulrich, my father and mother, taken at the farmhouse where my mother lived at the time of their wedding.

I Am Human Conflict and Adaptation

Reprinted from Behaviorists for Social Action Journal, *Vol. 3, 1981*

In 1949 I left my home in the small village of Eureka, Illinois, to attend North Central College near Chicago, and except for two years on a Navy tanker as a sailor, I've been involved with universities ever since as a student, teacher and explorer of human behavior.

Eureka Is Always Changing

I remember
                                  a season
                                                                    long past
When
                                  light-hearted
                                                                   but often confused,
I laughed
                                  most of
                                                                      the while.

Slowly I came to feel that I'd been trapped in a sheltered valley,
Surrounded by mountains of dogma.

    Then came
                the chance
                              to run
  Oh, and
                a long hard
                              run that was.
  And often
                when no one
                              was looking
          I would stop a while to cry,

For no matter where I went, there seemed to be
    no more sheltered valleys
And the sun was always
Just up ahead.

That was
                very, very
                              long ago.
  When I ran
                to find
                              the light
  And learned that
                Eureka is
                              always changing.

Archimedes, upon making an important discovery, shouted "Eureka!" which in Greek means "I have found it." During my years of exploration, I have often felt that I had found it…but the feeling was always short-lived and before long I would once again have to depart on still another quest and learn again that the quest is never over. I should like to emphasize here that my research and exploration have not been confined to the university laboratory and to books, but have taken me many times to many nations and to all but one of the 50 states…and more important, in all my quests for a better understanding of the extremely elusive nature of our subject, i.e., humankind, I, as a behavioral scientist, have at all times been as much the subject as I was the researcher and experimenter.

Also I feel that it is important to share with you the fact that my travels have not been limited to the physical highways and byways of behavioral interactions, but also include the experiences of expanded consciousness and non-ordinary reality that are produced by ingesting various combinations of natural herbs and chemicals which dramatically alter one's thinking and perception of the universe of which we are all a part.

I said earlier that I had written of my conviction that I did not have answers to the questions which this conference seems to address. Still, you invited me…and I replied that I would be your guest. I will now attempt as honestly as I know how to relate to you my recollections of some of the observations I have made during my exploratory journey on the road of life and some of the questions with which I am still left as pertains to the subject of "Human Conflict and Adaptation in the City."

In the movie *Understanding Aggression,* people saw me as I peered out from the screen and heard me say that "this information on aggression and its causes must reach those people in control, that's every one of us. Every individual has some control over other individuals…but any change has to start at an individual level. We have to start teaching and living nonviolence at all levels." Now I spoke those words as an allegedly knowledgeable behavioral scientist, as a research professor from a large Midwestern university who at various times had received support from research grants from local, state and federal sources, including the National Institute of Mental Health, the National Science Foundation, the office of Naval Research and the Michigan Department of Mental Health, as well as funds from private endowments and foundations. Furthermore, I was a member of the appropriate scientific establishments and frequently jetted around the world to tell people of my findings…"We have to start teaching and living nonviolence," I said. Through the modern technology of the movie screen, I implied that mankind should live nonviolent lives. "Although singular aggressive acts, such as dropping a bomb, are destructive, we must realize that our own daily actions are ultimately doing more damage than any particular bomb ever did." I perhaps could have added "do unto others as you would have them do unto you," or whenever a man strikes you on one cheek, turn to him also the other, or love your enemy as yourself and do good unto them who persecute you. I might have said those things…but I didn't. I said, "We have to start living nonviolence at all levels" from a movie screen amidst laboratory equipment, flashing lights and the jargon of modern Western scientific and technological materialism.

My forefathers, as non-resistant Mennonite-Amish Brethren, had dedicated their lives to the principles of nonviolence. That's why they came to America. I, like many the world over who have tasted the delicious fruits of materialistic grandeur and affluence, had forgotten my spiritual roots. The old religious principles were merely superstition, science was now in command, and salvation through its teaching and finding was what I was pushing. What had happened, although I didn't realize it at the time, is that we had simply given up one religion for another. To me, Christianity had become oppressive, and I observed many Christians often living one way and acting another. Indeed, the frequently-evil role of the Christian church in relation to the brutal and callous treatment of the North American Indian and the African Black slaves often made me ashamed of my Christian connections. The Christians were dogmatic: only they had the truth and only their words were right, and if you disagreed with them too vigorously, they might kill you. So I ran to embrace another "ism" only to have drift to my consciousness that it had, in many instances, come to the same position. Radical behaviorism, humanistic psychology and their leading proponents had become as rigidly dogmatic as many of my Mennonite Brethren and as equally adept at saying one thing with their mouths and something just the opposite with their other behavior…and worst of all I was as guilty as any. That's always such an embarrassing discovery. It's frustrating. It makes you angry and aggressive.

And the beat goes on… Dear Dr. Ulrich, would you be so kind as to come to Mexico City and participate in a conference on "Human Conflict and Adaptation in the City"?

Dear Dr. Jose Remus Araico: Sir! I will come to Mexico City and participate in the human conflict of your city and I will adapt there as I do in Detroit, Chicago, St. Louis and all points in between, and I will talk of the need to start teaching and living nonviolence in words only slightly altered from those I heard Sunday after Sunday on the prairies of Illinois from my fellow Brethren of the Mennonite Church, who taught me about the need to practice nonviolence.

> What has been is what will be
> and what has been done is what
> will be done. And there is nothing
> new under the sun.
> —Ecclesiastes 1:9

The Coming Down

I chose to drive to Mexico City for this conference, for it seemed to me to be relatively less violent than traveling by jet. As I indicated earlier, I wrestled for a long time over the decision of whether or not to attend...because I truly feel that it is probably the case that my countrymen and I have nothing to offer you by way of solution to your problems...indeed more often than not we are the problem. We are a nation of addicts. Human beings trapped into the most horrendous form of addiction...that of over-consuming absolutely everything we touch...and we, the rich of every nation, whether they call themselves first, second or third world countries, are teaching our addictions to others. We are pushers of overconsumption and bigness. Furthermore, we as the addicted rich will use any form of violence necessary to get the energy needed to maintain our habits, even if it means further starving the already oppressed, whether they be the poor of Colombia or India or the United States. Wherever the materials and goods can be found which are necessary to maintain our habits, we will go and take them.

I said at the onset that I have no solutions to the problems I perceive to be addressed by this conference...I bring only myself, my experiences as an explorer, some of my friends, and the questions which remain unanswered.

To clarify this, let me take you back for a while and a ways into my personal history. I told you at the beginning about my father and how he had to leave the farm in the middle 1920s. My mother also grew up on a farm and did the work that living on a farm required...and their fathers and mothers before them, they too were of the soil...of the earth. Hunters, fishers and farmers... Anabaptist emigrants from Europe who fled their homeland in order to be able to live their nonviolent beliefs. Their beginnings as a communal group were in Switzerland in Reformation times. They were called Mennonites after a Dutch priest named Menno Simons, but their founding was actually accomplished by followers of Ulrich Zwingli, and their first name was simply Brethren. The enemies of the Brethren called them Anabaptists because the Brethren refused to accept infant baptism. The actual birthplace then of the Mennonites or Brethren was the city of Zurich, Switzerland, in the year 1525. The city council of Zurich had decided to suppress the small company of counterrevolutionaries who refused to have their children baptized. *Although the reasons may differ, human conflict and adaptation in the city was a problem then just as it is today in Detroit and Mexico City.*

As is so often the case, suppression didn't totally work, although it severely impaired the movement. Eventually, because of such suppressions, the process which started in the city moved out into the mountains and farmlands, taking hold among the peasants who were already oppressed. Before long, however, the persecution followed the movement whenever it could be found and eventually succeeded in driving the Brethren from one country to another. Under the leadership of Jacob Hutter and others, one branch of the church in Austria in 1528 had adopted Christian communism as a way of life (the Hutterian Brethren) and were able to maintain themselves for a long time in their native land, but they also were finally driven out. By the year 1770, the last remnant of this group had been forced to flee to Russia, from whence they all migrated to South Dakota in the years 1874-1880. Descendents of this group are still living in community settlements known as "Bruderhofs" in South Dakota, U.S.A., and Manitoba and Alberta, Canada.

Given a historical perspective and the freedom given to different forms of worship today, it seems strange that such a severe persecution fell upon the shoulders of these early Anabaptists. Their stand against infant baptism, however, was not their only downfall. The most critical point related to their belief that children of God have to start teaching and living nonviolence at all levels. I said the same thing in 1970 with the help of the modern technology of filmmaking. My Amish-Mennonite Brethren ancestors said it and tried to live it beginning in 1525. They had read the New Testament and gathered that another way of life was necessary, and they began to witness through their lives that they indeed intended to be nonviolent. The essence of this Anabaptist revolution was not to overthrow the existing state, but simply not to conform to the ways of the world. Since the state would not tolerate the desired practices of the Brethren, the Brethren chose what for them seemed the only way out, i.e., a separation from the existing society into their own communities. Furthermore, they believed that the decision to join the fellowship should not be forced at birth via infant baptism, but that a person should voluntarily make this decision. In certain respects this "separating one's self from the existing society" is similar to the communal movement which has occurred at various times in history and most recently during the late 1960s. I and the people who traveled with me on this journey to Mexico City live on such a commune on the outskirts of Kalamazoo, Michigan. It's called Lake Village and it represents in a very basic way our own personal attempt to answer some of the problems of human conflict seen in city life, which often seems based upon the need to acquire more material wealth as opposed to de-escalating our material needs and practicing a more spiritually-based sharing.

These modern day efforts are indeed reminiscent of another cornerstone of the Anabaptist movement, which was the insistence on the practice of true brotherhood and love among members, which demanded the actual practice of sharing possessions to meet the needs of one another. Still the most important issue remaining as the basic foundation of the Amish-Mennonite Brethren movement was the principle of peace, love and nonresistance as applied to all human relationships. Not only were these people dedicated to a life of nonviolence toward others; they were committed to not resisting violence toward themselves. The Brethren understood this to mean complete abandonment of war, violence and other forms of conflict, particularly the taking of human life or causing human suffering. Here is what some of the early leaders were saying:

Conrad Grebel said in 1524: "True Christians use neither the worldly sword nor engage in war since among them taking human life has ceased entirely."
Pilgrim Marpeck in 1644: "All bodily, worldly carnal, earthly fighting, conflicts and wars are annulled and abolished."
Menno Simons wrote: "The regenerated do not go to war, nor engage in strife. They are the children of peace who have beaten their swords into plowshares and their spears into pruning hooks, and know of no war."

Principles such as these formed the basis of this early movement. It should be emphasized that it was these same principles which the movie, in its modern Western-world way, in a sense rediscovered via the methods of behavioral laboratory science and experimentation. Sincerely, the modern communal movement of which we at the Lake Village commune in Kalamazoo, Michigan, are a part is also characterized by some of the same precepts. My own early forefathers thoroughly believed and resolutely practiced these beliefs. As you may recall, among other things I said I served two years as a sailor on a tanker. I was a member of the United States military establishment and as such trained to do violence unto whoever happened to be the United States' current enemy. During the Second World War, there were other descendants of these early Mennonite Brethren who also had obviously drifted from the position of alleged nonviolence and nonresistance to one that obviously once again allowed for the taking of another man's life. What is it that makes us change from time to time, from one position to another? Once again, in order to illustrate this transition, allow me to reach back into the data of my own hereditary and environmental evolution.

In many respects, my forefathers could be looked upon as creative leaders far ahead of their times. Certainly they had more credibility than I did when I said, "We must begin to teach and live nonviolence at all levels." They said it with their lives and were often put to death. Yet when I said it in the movie in the early 1970s and before, I was really never persecuted. Only when you begin to really live a belief that conflicts with current popular practice will the persecution come. I was saying it, but in a whisper that posed no threat to the established controlling powers.

My ancestors were saying it and living it (i.e., refusing to participate in wars) in a day when both Catholic and Protestant churches endorsed war as an instrument of political affairs and used it in religious combat. Both the Protestant and the Catholic churches then did not hesitate to kill individual dissenters. Enormous numbers were thus slaughtered, since it was comparatively easy to determine who was an Anabaptist by the simple expedient of asking him point-blank: . . . . An Anabaptist disdained to save his life by telling an untruth and concealing his identity. When I finally completed the movie which you saw earlier, it was decided by the Office of Naval Research personnel, and agreed to by me, to not mention in the credits that United States Navy Research funds were used to help finance the film; it was felt that its anti-war stance might cause repercussions in the Navy and the United States Congress. I have noticed on many occasions that the morality of scientists and our alleged commitment to be honest at all costs about the data is a very flexible value.

Many of the early Anabaptists, however, refused to compromise. On Christmas day, 1531, an imperial provost drove 17 men and women into a farmhouse near Aalen in Wurttemberg and burned the building. Three hundred and fifty Brethren were executed in the Palatinate before 1530. At Ensisheim, the "slaughterhouse of Alsace," 600 were killed in a few years. In the small town of Kitzbuhel in the Tyrol, 68 were executed in one year. Two hundred and ten were burned at the stake in the valley of the Inn River in Austria. In Holland, at least 1,500 were executed. The Hutterian Chronicle records the execution of over 2,000. Before long, it became obvious that survival demanded that the Anabaptists move. A brief account of one of my great-great grandfathers, Christian Reeser, is typical of the exodus. He was born in 1819 in the province Lorraine in a little village called Gavanhusan, three miles from the larger village Sarreguemines, east of the city Metz. His father and grandfather before him were Mennonites who had come to Alsace-Lorraine via the Palatinate from Switzerland. With the coming into power of Napoleon after 1799, conscription became a common practice in most European countries.

Although religious tolerance had greatly improved compared to that of several generations before, there was no assurance that it would continue. And while they were persuaded among themselves to endure any possible suffering that might come to them, Christian's family also felt there was nothing to prohibit them from leaving for America, where the individual conscience and individual freedom were allegedly respected. The brutal and savage persecution of the American Indian that was taking place at this same time in North America, which in a sense made room for people such as my nonviolent, nonresistant ancestors, was a paradox of which my great-great grandfather, as well as many others, seemed unaware.

Indeed, how my family looked at the Indian and his treatment during those years is for the most part not clear, except that they seemed to be unaware of what was happening. When they came to this country, they were following their religious beliefs, for the most part isolated from the politics of expansionism which produced the Indian wars. Settlement was generally made on lands already stolen from the tribes who were consistently being pushed west. It is the case for many North Americans that the atrocities perpetrated against the Indian were not known and, when they were, it was done in such a way that it made the American Indian always look as if he were the savage rather than the European invaders.

It may very well be that there was no direct violence from the Anabaptist immigrants toward the Indian. But the indirect violent effects upon the Indian nation of the takeover of the whole North American continent shows that violence can be very indirectly manifested. Only recently have books such as *Bury My Heart at Wounded Knee* and *Custer Died for Your Sins* reached the consciousness of some North Americans. Books such as these detail the atrocities perpetrated by the white community against the generally less aggressive native Indians right up to the present times.

In a story written by a cousin in regard to another great-great grandfather and grandmother, there is mention of an incident concerning an atrocity by U.S. Army Rangers against a peaceful Indian tribe.

The following is a brief account of an event which occurred near my cousin's home in the middle 1800s.

Friends:

To the best of our knowledge and belief Peter Gingerich and Magadalena Nafgeger Gingerich his wife are the first known generation in our family history. They established their home about 5 miles north of Metamora in Metamora Township. The property was originally all on the west side of the road. A search of the records in the Metamora Courthouse confirms the location. A plot showing 200 acres of land in Section 5 with title in the name of P. Gingerich was found.

Chester Schertz, a great grandson of the Gingerichs, lives near the ancestral home and is familiar with the area. He is reasonably certain that they migrated from the Alsace-Lorraine Provinces but he has no proof of it.

When the early settlers came to this area in search of homesites, they sought out farmland where ponds or creeks were located. Wells dug during this period of our history were shallow. All creeks and streams in the area drained into the Mackinaw River. Chester said there was a mill pond on this farm where feed was ground and there was also a log cabin on the back of the property.

He doesn't know who lived there but it is possible that the Gingerichs might have lived there during the earlier years. The Gingerichs lived near the border of Partridge Township. There was an Indian village in this area. The village was located between Partridge and Richland creeks and consisted of 30 to 40 wigwams. It was considered a prosperous village as the wigwams were larger than the average and were substantially built. These were the Potawatomi Indians, a friendly tribe. While their Chief Black Partridge was away from the village on a Mission of mercy and the hunters were also absent, rangers came into the village. They massacred 30 defenseless old Indian men, women and children and completely destroyed their village. A monument has been erected in the Spring Bay area as a Memorial.

The earliest roads in the county were probably marked by dragging a log through the tall prairie grass, as time passed they became wagon trails and eventually dirt roads. A road was built from Chicago to St. Louis—some changes were made on this road later. There was also a Santa Fe Stage Coach trail built which crossed the north-south road connecting Chicago and the west coast.

Although the account appears generally sympathetic to the Indian plight, no great feelings seem to be expressed one way or the other as the story goes on to tell how the earliest roads got started.

The fact is the North American Indian was as spiritual and in many ways as nonviolent, before the coming of the white man, as were the Anabaptist Brethren represented by my early relatives. Furthermore, the Amish-Mennonite Brethren doctrine of nonconformity to the world with primary allegiance to a higher spiritual power has been fervently adhered to by both the Amish branch of the existing Anabaptist movement and the traditional American Indians.

> When the white man invaded our land it was an act of oppression. Now this oppression extends more and more to non-Indians as well, to minority groups, to people in underdeveloped lands, to people of new generations with new ideas, to all the people outside the government establishment.
>
> But we traditional Indians don't participate in that system. We're oppressed by it, but we don't try to be a part of it. You can't go to another people's land and try to kick everyone there off the land when they have nowhere to go, and kill most of them in the process, and then say that the ones who are left are supposed to join your club. That's wrong. We don't like their club and we won't join it. If it were a good club they wouldn't expect us to, and they'd leave our club alone. And they would leave other peoples in other countries alone. Everyone has his own club. If it's a bad club, it's no one else's business. The people will learn in their own way. No good system tries to spread itself. It's good to help people, but it's wrong to spread systems. It's wrong to spread beliefs. It doesn't matter whether it's Christianity or what it is, or whether it's supposed to be the best belief in the world—and there is no such thing—it should be told only to those who ask. It's wrong to spread any ideology by intimidation, and that means Christianity, communism, capitalism, democracy or anything else.[1]

---

1. Doug Boyd, *Rolling Thunder,* New York, 1974, pp. 39-40.

These words of the Cherokee Indian medicine man Rolling Thunder come as close as any to being a theme of conduct in relation to the issues of human conflict which all of us might well consider and perhaps strive to follow. My ancestors found something good and believed in it and were persecuted for it. They then had to flee for their lives to another land, where they found a form of freedom...which unfortunately was built from the spoils of another human loss...and then we tried to make the Indian a Christian!

Think again as to why you invite North Americans to your conferences on Human Conflict and Adaptation in the Cities. What answers are you seeking?

In 1967 I was invited to the capital of Michigan by an assistant to the Governor to give my views on the race riots that had occurred some weeks before. There was concern in the "halls of power" that riots might soon be spreading and they were interested in *how to control them.* Is it the case that there are those who wish to control human conflict in the city? I personally find it impossible to control the conflict I find within myself. Current and ancient history suggests that a similar state of conflict to that of mine exists in all men. Maybe it will become clearer as the conference continues as to what my role was intended to be and what it is we are seeking here together.

From the Farms to "the Cities" the Conflict Remains

And so it was in 1838 or 1839, when Christian Reeser was 19 or 20, that he and his three brothers and their sister set out on foot from southern France for the port of Le Havre, where they boarded a sailboat which was scheduled to land in Canada but because of severe storms drifted as far south as the equator, finally to dock in New Orleans, Louisiana. From there they headed toward Butler County, Ohio, from where another great-great grandfather, Christian Schmidt, who had also come from Alsace-Lorraine, was about to leave for Illinois.

The 1830s were a long way from the 1930s, however, and to me these stories of the past were as yet barely audible. They certainly had little direct effect on the life of a small-town lad whose major concern was playing baseball with my friends, some of whom were not Mennonite, and thus the beginning of new and different ways of looking at life. My father had left the farm: I was a TOWN KID!.... And beginning my earliest lessons on human conflict and adaptation in the city.

For some reason or another, my father and some of his brothers fit less comfortably into the traditional Mennonite way of life. I refer to them now as Mennonite because that's what we were called, having somewhere dropped the Amish Brethren and Anabaptist labels. Certainly the Depression had its effect. Mennonite farmers, like farmers who were not Mennonite, were losing all they had of material wealth, yet the abundance of natural resources which the Midwestern United States offered made material wealth "easy come, easy go." My father, who by now was selling and working on tractors and other farm machinery, was not as much influenced by the tight church authority that had earlier come to govern the lives of the sons of the Anabaptist revolutionaries, whom the Protestant Reformation had spawned, martyred, then forced to be roving refugees. There was somewhat more stability now. Although my father had made the move from the farm to the village, it was a small village and he seemed to be able to handle the conflicts. My mother was more shy and still more a part of the conservative Mennonite tradition. She was a pacifist in her verbal behavior, while my father was not. My mother was more inclined to remain aloof from the ways of the world; my father, again the reactionary, was willing to take anything the world had to offer. My father, it seemed, was simply tired of taking the harassment that apparently went along with being a nonresistant pacifist Mennonite in a materialistic world, such as the United States with all its wealth.

Indeed, many of the church members were doing very well and now only talk like their forefathers while behaving like any other money-hungry gringo. My father was extremely verbal and outrageously direct. If he didn't like the sermon at church, he would walk out or lean his head on the pew in front of him and simply go to sleep. He smoked, drank a little and was more than willing to swear when he felt the occasion called for it. Once a fellow church member made a deal to trade cars with my father and then switched some older tires with the better ones that had been on the car at the time of the trade. My father went after him and, in front of a large group, informed him that he was very wrong in his actions. He used forceful words that I'm sure were not what Anabaptists had in mind in the early 1500s as a nonviolent nonresistant stance. I'm sure they also couldn't conceive of tire-switching by one of the Brethren.

Times of course change and the conditions that brought the Mennonites to the United States, first to its farms and then to its cities, were bound to change them—just as conditions had changed people in the past, producing the need for another reaffirmation, reformation, revolution or whatever you wish to term such social upheavals.

In any event, the conflicts that began to form about me were accelerating. The following is a brief portion of another paper in which I tried to explain my dilemma.

Some Confessions of an Amish Warrior

For as long as I can remember, people have acted in ways which suggested to me that they were afraid of obligations to act against their own "free will." When I was a child, growing up as a Mennonite in Eureka, Illinois, it was the devil that needed to be held in check. That old bastard could trap your "free will" into doing a lot of dumb things.

Now let me tell you quickly about Mennonites. Basically, we were very religious farmers who stuck closely together in semi-communal arrangements centered about the church. Since our founding in 1525, we ran from country to country persecuted for our strong belief that we should not kill our neighbors. We were pacifists who somehow managed to "anger" our European neighbors so dramatically that they systematically set out to kill us. Since 1525, we have argued ourselves into hundreds of splinter groups that include diversities all the way from Shipshawana Amishman to a President Eisenhower Brethren.

As a Mennonite, I found that there was right and there was wrong. We were free, yet controlled. There was pleasure and there was pain. There was reward and there was punishment…sometimes given and sometimes earned. There was guilt…sometimes deserved and sometimes not. There was also a fair amount of confusion, much like there is today. However, I am now a Ph.D. psychologist and supposedly exempt from childish confusion.

In 1941, when I was ten, there were Japs, Nazis and some "misguided" Wops who "all got together" and "determined" that "they" were going to conspire so as to "alter the freedom" of the "good-guy allies." Sometimes, however, given the efficiency of our silver-tongued propaganda machine, kids with names like Stromberger, Webber and Ulrich (the "Nazis"), and kids with names like Micaletti, Gerardi and Pannone (the "dago-wops"), got mixed up as to whether they were the "good guys" or the "bad guys." They subsequently felt some need to design counter-control strategies to facilitate their own defense and survival.

There were numerous "ethnic" family groups whose hearts ached with confusion, due to hereditary and environmental events far beyond their control. The news media delivered daily imports on the events of the war, describing the friends and kin who still lived in what was once their home countries as evil murdering bastards.

In those days, I really didn't understand one damn thing about the eternal debate over the question of whether we are free or determined. I just went along, like Leonard Cohen's "Bird on a wire…and his drunken bum in the midnight choir," trying in my way to be free…and to get others to do what I wanted.

In 1941, 1 was still uncertain whether or not I had the ability, the responsibility or the freedom to choose right from wrong. But I did have a good, healthy, sick fear that most often I was "fucking up." (I was not using this term much in 1941. It became more commonplace with me about a decade later when I served in the U.S. Navy. The feeling remains the same, however.)

When you are ten years old and your country is at war with a country whose native tongue is the same one that some of your relatives speak and your name is frequently the same as those of enemy generals…and when you are known to be a member of an "off-beat" religious group who openly refuses to participate in another "war-to-end-all-wars"…and when other kids sometimes call you a yellow-bellied, C. O. Nazi, pacifist, chicken son-of-a-bitch…and when a church you often attend has rocks and yellow paint thrown at it, you might find yourself frequently getting into fights.

Now, allow me to explain how things became even more complicated. I was a Mennonite "warrior," a member of a minority within a minority. Sunday after Sunday, this minority received intense pacifistic propaganda designed to modify or maintain the behavior of all Mennonites in the anti-military effort. Yet my draft-age friends, some cousins and uncles on my father's side of the family, refused to follow the Mennonite doctrine of non-resistance and went to fight for freedom.

I stayed home and fought. Furthermore, I did not always wait for opportunities for defensive action. I often acted on the offensive and, whenever possible, kicked the shit out of youthful companions whom I was fairly sure I could whip and who might encourage future defensive actions. In a sense, I became a "Why Wait Pacifist." I virtually learned to hurt my brother before he could hurt me.

"Avoidance...Murray Sidman...punishment...operant behavior...Skinner." Without realizing the professional labels, I was locked into a series of escalating aversive encounters which involved some very basic competitive and aggressive human emotions. When I was hurt too badly, I hit!

In order to escape pain and to experience pleasure, I had to conquer my surroundings...which included other human beings. I was learning how (or so I thought then) to control the actions of other men in ways that would make me happy. I assumed that I was, of course, free to do that. The seeds of my interest and eventual entrance into the ranks of the behavior shapers, the human behavior modifiers, the behavior engineers or whatever synonym for "Boss" you wish to use, were being fertilized in the farmlands of the Midwest with powerful consequences. Through seemingly successful personal action, I gained more and more "control" over my life. The consequences of further positive reinforcement for those actions seemed to allow me to determine the course of my existence. According to some humanists, I suppose I was becoming self-actualized.

Success, however, can also facilitate failure. In this case, the early failure was a lack of astuteness in perceiving the core of the control vs. freedom myth. I, *who didn't even know who or what I was, was being sucked into the trap of actually believing that I was a prime mover, as opposed to simply being part of the show which could absolutely not be considered without reference to other events. Such events, when considered as a whole, would become the prime mover.*

In 1949, prior to the Korean War, I finished high school. In college, my primary interests were sports and winning at whatever I attempted. Eventually, like some of my uncles and cousins, I went into the service and became another Ulrich turncoat to the sacred cause of Anabaptist pacifism. For approximately two years, I toured the world as a deck-ape on the USS PAWCATUCK.

Big Is Not Necessarily Beautiful

The Navy, of course, allowed me or forced me to become even more worldly in Mennonite terms. I had already finished college and knew the ways of the city and world travel. I was an athlete and a vigorous competitor. I was certainly not nonresistant. Being small seemed to attract attempts on the part of others to test their manliness, and so long as I can remember I had people pounding on me. More often than not I found that people would keep it up until you hurt them or at least made them think you would. I had often been knocked unconscious in struggles of one sort or another and had adapted to many forms of conflict that for many would have been more painful than it necessarily was for me. Furthermore, I was a member of a culture that constantly taught us that we could do more, get more, go further and faster and higher! To get high you used morphine, or cocaine, or speed, or quaaludes plus fast cars, fast women, fast horses, fast planes, T.V. and flashy suits of clothes. If you're Black, a hippie or some other kind of *poor* nonconformist, they call you a junkie. If you are rich and can afford the same indulgence, they call you Mr. and maybe make you a Hollywood star or perhaps even a research professor.

What is the point! I came from a family not all that much unlike the poor peasants from third world countries that stream into urban areas like Bogota or Mexico City to escape the poverty inflicted by rural rape. Let me give an example of what might be considered "rural rape."

> Throughout the hemisphere, food production has fallen increasingly into the hands of modern "agribusinesses," the soya producers of southern Brazil, the wheat producers of northwest Mexico, that use the best machinery and techniques and as little manual labor as possible. In many cases, the huge United States farming companies have sold their properties, preferring to buy the primary products from nationals and take their profits in processing and marketing. With high birth rates and little job creation in the countryside, the rural populations have been spilling over into Latin America's cities. Between 1960 and 1975, the rural population of Latin America grew from 101 million to only 115.5 million, while the urban population grew from 98.7 million to 186.9 million. Stated differently, the rural share of the total population of the region fell from 50.6 percent to 38 percent in 15 years.

In a sense, the tender loving care that a human close to the land extends to it has decreased by the same 38 percent. Huge farming companies have not shown the willingness to put back into the land what they have taken from it. They don't love the land as much as they do the short-term profits they extract from it. That is to me "rural rape" and very much akin to what happens to the poor prostitute who is forced to submit in order to live. When an imbalance of energy occurs over too long a period of time, the conflict it produces no longer remains hidden. That is what you may be experiencing in Mexico City today.

Rape anything too often and it will die, but it will strike back in anger before it finally succumbs.

What is happening here is not unlike what happened to the Black who came from the farms of Mississippi or Alabama to Detroit and Chicago, or the Italian farmer from south Italy who went to Milan to make cars or to the United States to build tires. From all of these groups, there may eventually be a few sons and daughters who finish at the university and who will perhaps someday end up talking somewhere about "Human Conflict and Adaptation in the City." But their credibility, like mine, will never equal that of the person who right now exists at the poverty level, exemplified by the angry peasants in northwest Mexico who for six decades since they "won" the 1910 revolution are still without land, or the southern Black still living in big city "slavery" a hundred years after the Civil War.

Things, of course, never happen all at once. It seems one never learns something new right away. Even when you get killed. I imagine it takes a little while to get used to it. At least that's what happened to me. In 1971, not long after the film you saw was completed, there was a group of us who moved onto a farm called "Lake Village," much like the ones inhabited by my early kin. As I already mentioned, we were a part of the great commune movement of the middle and late 1960s. We were beginning to comprehend and appreciate the damage that technologically-addicted mankind was fostering on Spaceship Earth. We read, among other things, *The Greening of America* by Charles Reich, *Limits to Growth* by Meadows, Meadows, Randers and Behrens, *The Closing Circle* by Barry Commoner, and Paul Ehrlich's *Population Bomb*. We were no longer as sure as we once were that we could change North American culture via established political routes and thus began to seek alternatives through intentional communities. It was not totally unlike the Anabaptist movement that I have spent so much time telling you about, although the points of difference with the ruling power structure were not the same and the consequences not as severe. No hippies that I know of were burned at the stake, although the consequences for some were as severe as any issued for other so-called crimes against society. There were those who did get shot or beat up by authorities protecting the North American way of life, and many landed in jail.

I was then into drugs for about five years. The drugs were mostly hallucinogenic in nature, although I was beginning to experiment more often with cocaine as well as a variety of other chemical compounds. Of course, like many other people, I was a heavy marijuana user.

I had been involved in a variety of ways with the civil rights movement and anti-Vietnam War movement. In brief I, like others, was being "current," doing what for North American counterculture was popular, things that were "in" and needed to be done. Life for us was free, energy was abundant, sex was all around and the trip was "pure far out" in spite of its oftentimes negative aspects. Then it happened, and the Lord of the universe took and showed me still another reality...death!

To make a story that took an eternity brief is very difficult, but allow me to try. One late afternoon before the Christmas of 1971, after mainlining three good-sized injections of excellent cocaine and swallowing some very good LSD, I was taken for the trip of my life through all of evolutionary history right to the brink of Hell, where I discovered that it is indeed true… there is one and I was about to enter it forever, which I did until finally I died. About 2½ hours later by the earth clock (absolutely an eternity by the one I was tuned to during the trip), I began to come back to life. As I implied above, nothing is learned all at once, and I am to this day and will be of course forever affected by that event. All in all, it was just another side trip in the eternal voyage… Another exploration which led to still another and then another. I do, however, feel that that event was a very, very important experience in my life. It broke my faith in worldly materialism. Although many other events contributed, I finally came to see that the answers to our social problems were not going to be forthcoming via our scientifically-based behavioral engineering technology. The scientists and technologists such as myself had too many of the problems that we were trying to solve in others dying as cancerous residual within ourselves. We were our own greatest problem… Slowly the emphasis of my research and personal efforts of problem solving began to focus in on myself…and time and time again I was unable to manage my own problem behaviors let alone those of all Mexico City.

They Would Not Conform to the World!

Very few of my kin and my sect (except for the Amish members of the Anabaptist tribe) today remain on the farms. They live in cities and villages, drive big cars, wear the finest clothes, inhabit fine homes trimmed in the latest style and have two or three television sets that bring the world into their front rooms, where the real violence of the Vietnam War or the Chicago,

Watts and Detroit riots mixes imperceptibly with the violent entertainment of the current season and old Ronald Reagan reruns. When I was little, I was seldom allowed to see a movie. When I was 45, some of my family wanted Ronald Reagan, the former movie star, for president. When I was little, I was admonished by the Mennonite elders to be a pacifist. When I grew older, I watched the elders of the Mennonite Church condemn the young with their Christ-like beards as they marched in protest against the war and emigrated to Canada rather than fight. Some people shoot up heroin; others shoot up the evening news. "Temperance in all things," said my Sunday school teacher as she pushed her religious addiction, which had long since lost much of its early meaning (1525 circa Friesland; Holland; Zurich, Switzerland, Menno Simons, Conrad Grebel).

Human conflict and violence throughout the United States is as North American as home-baked apple pie. We are addicted to the sugar in both. It's part of our diet. We don't even notice it. Overweight Mennonites, like other North Americans, continue to wail against the evils of the world, with its subversive elements, Communists, SLA, pot or whatever...while all around us the trees are coming down and cement is being poured over another hundred thousand acres of once rolling prairie for still another six-lane highway, parking lot, jet port or drug factory.

You invite this North American son of the Anabaptist revolution to Mexico City...was there a reason? As you may know, I come from a land which carries roughly 5.6 percent of the world's population. We require, it has been estimated, 40 percent of the world's primary resources to keep us going. Do you like to look at junkies?

We are, however, beginning to worry about the future and how we will be able to continue supporting our habits. There are those who study raising better homegrown pot so the Colombians, Jamaicans, Mexicans and other suppliers will no longer hold us in their grips. And there are those who are studying ways in which nuclear energy will make us free of the swarthy Arab—in spite of the fact that large-scale nuclear fission is undoubtedly a profoundly dangerous change in the nature of our environment. The burden of proof is placed on those of us who take an "ecological viewpoint" as opposed to the ones who suggest that peaceful nuclear energy will not cause problems. Certainly the dangers of the atomic bombs have been made clear. However, its peaceful use (which may in the long run be even more dangerous than the bombs) is advertised via TV and every other media to the children in nonviolent, nonresistant Mennonite homes, as well as others, as being almost if not more desirable than Miss America! What is not said, however, is that once having created radioactive elements, there is nothing that can be done to reduce their radioactivity. We don't picture on our TV atomic energy-type radiation particles ripping like bullets into the bodies of old ladies, small babies and puppies. Yet there is no known place on Earth that can be shown to be safe for storing radioactive waste products created by nuclear reaction.

Wise men once thought that such waste would be safe in the deepest point of the ocean. However, Russian deep-sea exploration showed that plankton, algae and many sea animals absorb these substances and, as one animal feeds upon another, the radioactive materials climb the ladder (up or down, whichever way you want to look at it) and return to man. They are talking about storing radioactive waste in abandoned salt mines in Alpena, Michigan. Governor Miliken is listening to the experts who think it will probably be okay. Our local newspaper feels we shouldn't do it unless we get paid a lot of money. Well, that's how it goes…anything for a Yankee dollar. Junkies will steal and rob to support their habits…sometimes even kill…future generations.

Did you invite me to this conference to see a murderer? A member of the big U.S. tribe? Sometimes as I live amidst human conflict I think for a while that I almost understand it…but before long that feeling passes and I'm confused again. Maybe then you invited me here to help me! You know confused junkies don't like being junkies, except when they're high, and that doesn't last forever. You must always drop as low as your high was high…and breaking habits is terribly hard. Understand that I'm not talking down to anyone. I know that junkies don't live just in the U.S. of A. Everyone in this group is an over-consuming materialist relative to someone else.

Many of you I am sure are aware of the Gringo who lies within us all.

Los Gringos Dialogue

Gringo American—
I am the Gringo
Who walks within you.
I who force you
to speak my language.
I who insist that you dance
to my national anthem.

I am the Gringo
> Who left his native land.
I who was afraid
> to kill my neighbor.
I who was afraid
> to die myself

I am the Gringo
> Who walks within you.

Gringo Latino—
I am the Gringo
> Who walks within you.
I who hate you
> With my father's riches.
I who ring the bell
> to summon the maid.

I am the Gringo
> Who pities the peasant.
I who left his native land
> to kill my neighbor
I who wanted the
> Andes Gold for myself.

I am the Gringo
> Who walks within you.

Gringo Americano—
You are the Gringo
> Who claims I murdered Allende
You who drive your Land Rovers
> With Dr. Che on the bumper
You who talk revolution at the University Nacional
> Before leaving for vacation in Europe.

You are the Gringo
> Who despises the tourist and speaks against the building of New Hiltons

You deplore the theft of Indian treasures
  With Indian jewels round your wrist and neck.
You who teach Barrio Children to fight Yankee oppression
  While applying to graduate schools in California and New York.

    You are the Gringo
      Who walks within me.

Gringo Latino—
  You are the Gringo
    Who speaks at my country's conferences.
  You who are the sacred cows of Imperialistic Academia
    With power in your hands to take more than you give.
  You who suck the resources from our third world
    And sell the waste back for profit.

  You are the Gringo
    Whose cast off T.V. teaches us your values.
  You who claim our political allegiance with dollars
     Backed by gold stolen from my country
  You whose language I learn
    So that my children may get rich.

    You are the Gringo
      Who walks within me.

    CHORUS

    *Gringo Latino
    and
    Gringo Americano—*
      We are the Gringos
        Who walk together
      We are the they
        Who forget "we"
      Are..."us"!

I've talked with you now for several hours. There are still many observations that I would like to share and questions of course still to be answered, but the time is growing short. For me, Mexico and the rest of Latin America means many things. It is the home of Don Juan, the Yaqui Indian who taught me through Carlos Castañeda more about non-ordinary reality. I have been helped in your countries by the people and the sacred mushroom to know myself more as a child of nature and not as an outside force destined to dominate and conquer it. I cannot battle nature without battling myself. It appears that the natural capital of the world may be running out. We have known that it could happen sooner or later, but like most things we assume it will not be us who feels the big sting! Lincoln Barnett in his book on Dr. Einstein and "his universe" explains the possible end and possible re-beginning in this way.

> The universe is thus progressing toward an ultimate "heat-death," or as it is technically defined, a condition of "maximum entropy." When the universe reaches this state some billions of years from now all the processes of nature will cease. All space will be at the same temperature. No energy can be used because all of it will be uniformly distributed through the cosmos. There will be no light, no life, no warmth—nothing but perpetual and irrevocable stagnation. Time itself will come to an end. For entropy points the direction of time. Entropy is the measure of randomness. When all system and order in the universe have vanished, when randomness is at its maximum, and entropy cannot be increased, when there no longer is any sequence of cause and there will be no direction to time—there will be no time. And there is no way of avoiding this destiny. For the fateful principle known as the Second Law of Thermodynamics, which stands today as the principal pillar of classical physics left intact by the march of science, proclaims that the fundamental processes of nature are irreversible. Nature moves just one way.
>
> There are a few contemporary theorists, however, who propose that somehow, somewhere beyond man's meager ken the universe may be rebuilding itself. In the light of Einstein's principle of the equivalence of mass and energy, it is possible to imagine the diffused radiation in space congealing once more into particles of matter— protons, neutrons, and electrons—which may then combine to form larger units, which in turn may be collected by their own gravitational influence into diffuse nebulae, stars, and, ultimately, galactic systems.

And thus the life cycle of the universe may be repeated for all eternity. Laboratory experiments have indeed demonstrated that photons of high-energy radiation, such as gamma rays, can, under certain conditions, interact with matter to produce pairs of electrons and positrons. Astronomers have also determined recently that atoms of the lighter elements, drifting in space—hydrogen, helium, oxygen, nitrogen, and carbon—may slowly coalesce into molecules and microscopic particles of dust and gas. And still more recently Dr. Fred L. Whipple of Harvard has described in his "Dust Cloud Hypothesis," published in 1948, how the rarefied cosmic dust that floats in interstellar space in quantities equal in mass to all the visible matter in the universe could in the course of a billion years condense and coagulate into stars. According to Whipple, these tiny dust particles, barely one fifty-thousandths of an inch in diameter, are blown together by the delicate pressure of starlight, just as the fine-spun tail of a comet is deflected away from the sun by the impact of solar photons. As the particles cohere, an aggregate is formed, then a cloudlet, and then a cloud. When the cloud attains gigantic proportions (i.e., when its diameter exceeds six trillion miles), its mass and density will be sufficient to set a new sequence of physical processes into operation. Gravity will cause the cloud to contract, and its contraction will cause its internal pressure and temperature to rise. Eventually, in the last white-hot stages of its collapse, it will begin to radiate as a star. Theory shows that our solar system might have evolved, in special circumstances, from such a process—our sun being the star in question and the various planets small cold by-products condensed from subsidiary cloudlets spiraling within the main cloud.

Presupposing the possibility of such events as these, one might arrive ultimately at the concept of a self-perpetuating pulsating universe, renewing its cycles of formation and dissolution, light and darkness, order and disorder, heat and cold, expansion and contraction, through never-ending eons of time.

At least this theorizing suggests to me that there is still much to be learned about all of nature, which again includes we humans as a part.

I came from a place, however, where it often seems that those in power feel as if man is the center of the universe and its prime mover, led by the high priests of science, medicine and law.

And often it is as if we are saying to our students and to the community at large…we know…we have the answers. It's almost as if some of the social scientists with whom I have been closely associated are saying that we do understand the laws of behavior sufficiently well so that wars and civil disturbances can never happen, human conflict can be solved. A recent advertisement for Dr. B. F. Skinner's book *Walden Two Reissued* had the following quotes:

> "Skinner points the way to a world free of the problems of pollution, over-population, resource shortages, nuclear war, economic distress. (He) suggests that human behavior management is a realistic and perhaps necessary solution to the problems threatening the world today."

Professor Skinner is a behavioral scientist whom I have admired for years. He has visited Lake Village commune, which in a sense received the impetus for its founding from his book *Walden Two*. Together we recently traveled to Twin Oaks, which declares itself as a bonafide Walden Two commune and is in my estimation one of the more innovative social experiments presently in progress in North America. Skinner's book *Walden Two* is credited with beginning Twin Oaks. Twin Oaks is not, however, a society in which human problems are solved by a scientific technology of human behavior management and in which many contemporary values are absolute, as the ad for Skinner's book *Walden Two* claims is pictured in the novel. Indeed in many ways the data of the Twin Oaks Walden Two experiment show that the materialistic values of scientific behaviorism, no matter how fervently held by either Skinner of the founders of Walden Two, have in no way overcome nor superseded the mystical bounds of spiritual brotherhood, which remain impervious to modern efforts to either measure or discount them.

Skinner and his disciples have not pointed the way to a world free of the problems of pollution, overpopulation, resource shortage, nuclear war and economic distress, as the advertisement by the Macmillan Publishing Company for Skinner's books suggests. As behaviorists whose beginnings stressed paying attention to observable phenomena, it should be obvious with even the most crude measuring devices that the excess in claims made by the Macmillan Company for the *Walden Two Reissued* book is at best a joke and at worst dishonest. Human behavior management has not proven itself to be realistic nor is it, as the ad proclaims further, a necessary solution to the problems threatening the world today. None of us in North America, including Professor Skinner, myself and all of our behaviorist colleagues, regardless of how firmly they remain in the radical behaviorism faith in scientific solutions, have pointed the way to a world free of problems. Just stop for a moment and look at us...look at our lives! Indeed, as I implied before, we are often obviously more a part of the problem than the solution. We at Lake Village have not solved our own small day-to-day conflicts within ourselves, let alone arrived at the point where we can, based upon some behavioral management data, solve yours. We forgot for a while that we were not gods, and although we claimed to be a part of a natural universe, we found ourselves more often communing with the supernatural world of television, big 747 planes, computers and the ways of the city...all the while losing more and more touch with the gentle Mother Earth which indeed supports us all. We support and are supported by publishing companies that are willing to promise too much more or less than the truth via advertising hypes. The advertisement by the Macmillan Company is designed to sell an old classic by Skinner as something new because a few additional pages have been added at the front of the original book. We are our own problem.

The power behind most human behavior management is still determined by the degree to which any one person is still hungry for the dollar, still addicted to the habits which years of conditioning put in us. Look at what we do, not what we say. There are many who believe that the best way to show one's sincerity about living in an America made up of small communities where people live productive and creative lives free of the pressures and violence of the big cities is to move out...not just talk about it on the TV circuit...but remember that won't be easy to do, and some of the world's most renowned behavioral engineers have not been able to do themselves as they would have others do. And I drove all the way here to come to one hell of a big city with one hell of a bunch of problems. Are my words consistent with my actions?...and are yours?

As we truly come to understand ourselves as children of nature, simply a natural part of the natural universe and not its maker, it is hard to imagine then that we stand ever near the brink of understanding nature, let alone its problems. Certainly we are not the saviors. Again hear the American Indian Rolling Thunder speak and listen to the implications of his words for our problem of human conflict.

"...This idea I've found in some modern people that there's no good or bad, that it's all the same, is pure nonsense. I know what they're trying to say but they don't understand it. Where we're at here is this life, with all our problems, there's good and there's bad, and they'd better know it.

"As long as so many people accept this modern-day competition, willing to profit at the cost of others and believing it's a good thing; as long as we continue this habit of exploitation, using other people and other life, using nature in selfish, unnatural ways; as long as we have hunters in these hills drinking whiskey and killing other life for entertainment, spiritual techniques and powers are potentially dangerous. The medicine men and traditional Indians who know many things know also that many things are not to be revealed at this time.

"The establishment people think they have a pretty advanced civilization here. Well, technically maybe they've done a lot, although we know of civilizations that have gone much further in the same direction. In most respects this is a pretty backward civilization. The establishment people seem completely incapable of learning some of the basic truths.

"The most basic principle of all is that of not harming others, and that includes all people and all life and all things. It means not controlling or manipulating others, not trying to manage their affairs. It means not going off to some other land and killing people over there—not for religion or politics or military exercises or any other excuse. No being has the right to harm or control any other being. No individual or government has the right to force others to join or participate in any group or system or to force others to go to school, to church or to war. Every being has the right to live his own life in his own way.

"Every being has an identity and a purpose. To live up to his purpose, every being has the power of self-control, and that's where spiritual power begins. When some of these fundamental things are learned, the time will be right for more to be revealed and spiritual power will come again to this land."

To some of my scientist colleagues I know I have ceased being a scientist and instead became a mystic, yet in my defense I reply that in fact I continue to report the data as I see it.

Einstein, whose philosophy of science has sometimes been criticized as materialistic, once said:

> "The most beautiful and most profound emotion we can experience is the sensation of the mystical. It is the power of all true science. He to whom this emotion is a stranger, who can no longer wonder and stand rapt in awe, is as good as dead. To know that what is impenetrable to us really exists, manifesting itself as the highest wisdom and the most radiant beauty which our dull faculties can comprehend only in their most primitive forms—this knowledge, this feeling is at the center of true religiousness."

I told you earlier about a time when one of my odysseys took me to Hell and showed me death. There have been other excursions, however, which allowed what may be called a chemical glimpse of paradise. Neither trip let me return, however, with the answer. More exactly, I was simply allowed to return. I have often thought of myself as both subject and experimenter. Were I a thespian rather than a scientist, I might see myself more as both spectator and actor on the stage of existence, or as an athlete both player and fan. No matter how I name it, I remain my own greatest mystery. I do not understand the universe in which I have been cast, for I do not understand myself. In spite of many years at the universities as a student of behavior, I still comprehend almost nothing of my most complex organic processes and perhaps even less of my capacity to perceive, of my capacity to reason and my capacity to dream. Like the physicist Heisenberg, my inescapable impasse is that I, as part of the world I seek to explore—my brain, my thoughts, my body—am a collage of the same elemental particles that make up those same drifting clouds of interstellar space that have been theorized as the essence of a self-perpetuating universe.

I came to Mexico City as a child of immigrants from another land, as are no doubt most of you. I have no answers to the question of human conflict and adaptation… I *am* "human conflict and adaptation." I once thought as a behaviorist that I and others like me could program the world so that humankind's social problems would diminish…then I discovered that such programming had to begin with the self. When we see problems around us, more than likely they are also within ourselves. If I have a destructive thought or wish it, it will have its effects on someone, if not on another then it will work back on me. Cities all over the world are in trouble and the world itself, like any other living organism that has been abused, can get sick. People have forgotten the laws of nature and the need for balance. All over the world the countrysides are being raped, and the energy of both the land and the people of the land are being exploited, often to the supposed benefit of those living in our modern cities. Governments which reside in the city (while relying on the countrysides for their true source of life) have developed into grotesque and overbearing powers which are causing disharmony and destruction all over the planet. Allegedly wise leaders continually fail to see the causal relationship between mental illness, crime and other social problems with air and water pollution, and the destruction of trees. We continue to rape the world's natural resources, including many of its people, so that more goods are available to buy and sell. When we lived close to the land and had an intimate relationship with the environment, we learned about the natural world. We were one with nature and one with ourselves. There are American Indians who tell us that we can learn about air pollution, the human condition, levels of anxiety and hostility, and about the conditions of the earth or the coming of earthquakes and floods, if one knows what to look for in birds and their habits as well as with related natural phenomena. As I have moved back to the living laboratory on the land at Lake Village from my laboratory in the city, I have come to realize that it is not the religion and spiritual belief of the Indian or the Amish Brethren which is supernatural.

 They and others of the land and forest and the water are the ones who are natural. The supernatural beliefs are most often found among the Western materialistic power establishment and their pseudo-natural science with its mutant technology. The skills of many people of the land who are truly behaviorists, familiar with nature and its most basic lawfulness, are to me more firmly based in the truly scientific method. These are the ones, as they interpret their complete and more accurate observations of life as it is in nonsupernatural settings, who will inherit what is ever left of this Earth. As I have returned to a closer relationship with the land and its animals and the water and the trees, I have often felt the oneness of the world which is so common when one is experiencing the non-ordinary reality which results when ingesting the sacred mushroom. We live in a universe that cannot be contained by boundaries conjured up by man; the earth, the cities, the country, different nations, water, air and people are intertwined…and at the moment, as this conference suggests, there seems to be an illness in the organism, this Earth, this struggling living being.

Let me close with a bit of Hopi Indian prophecy which pertains to an approaching change that is called by some "the day of purification." This prophecy correlates with that of ecologists who believe that the imbalance in nature has passed the point of no return. Yet the traditional Indian does not await some ecological doomsday but instead anticipates the moment of climax as a sort of time of healing. The conflict of the city is a symptom of the larger illness.

When you have pollution in one place it spreads all over. The Earth is sick now; it's been mistreated and the cities are the sorest spots. The natural disasters which are now occurring and will continue to occur are the natural readjustments that have to take place to throw off the sickness.

Rolling Thunder says that we and the Earth together are a living organism. Together we form the body of a higher being who, in consort with us, has a will and wants to be well and who at times, also like us, is less healthy and more healthy both physically and mentally. As I said in the movie and have tried to say throughout this visit, the first step to any solution lies within ourselves. We must treat our own bodies with respect, and the same is true of the Earth, with all its varied nature. When we harm it, we harm ourselves. Let us, therefore, join together in a silent resolve to care for the spirit and the body of the cities and lands in which we reside as we would have our own spirits and bodies cared for in return.

Foreword from *Anabaptists in America*

For those of us whose roots stem from the Anabaptist movement which dates back to the early 1500s, there is a bond which holds us within a cultural community that in a sense makes us all family. Whenever we travel we see the tribal signals…a train station in Chicago with a large group of Amish sitting together waiting to return to Ohio—there is a family to whom somewhere, somehow, we are related. "What's your name?" I ask the dark-eyed 12-year-old, too shy to answer, who turns to her brother for help replying to this strange English questioner. "Mary," comes the reply. "What's Mary's last name?" "Yoder," he says…or in a medical center in Kalamazoo, Michigan, waiting for your name to be called for your annual check-up, you hear the name "Esther Troyer," and you watch to see who stands and again have a sense of relationship as you see a lady rise to whom, somehow without asking, you know that were you to inquire you would find, before long, you were related.

I am a part of that heritage. My cousins, my uncles and aunts, my grandparents great and greater have names like mine and are linked spiritually and physically to folks called Schertz, Yoder, Garber, Schrock, Imhoff, Wagner, Householder, Oyer, Miller, King, Zehr, Knapp, Baer, Litwiller, Harnish, Neuhauser, Yordy, Zimmerman, and so it goes.

My father left the farm in the middle 1920s to find work in the nearby town of Eureka, Illinois. My mother also grew up on a farm and did the work that living there required, and also worked in Eureka following graduation from high school. Their fathers and mothers before them were also of the soil, of the earth, hunters, fishers and farmers—Anabaptist emigrants from Europe who fled their homeland in order to come to America and live their nonviolent beliefs.

Their beginnings as a communal group can be traced to Switzerland during the Reformation. They were called Mennonites after a Dutch priest named Menno Simons, but their founding was actually accomplished by followers of Ulrich Zwingli, and the first name they called themselves was simply "Brethren." The enemies of these brethren called them Anabaptists because they refused to accept infant baptism. The actual birthplace of the Mennonites, or Brethren, was the city of Zurich, Switzerland, in the year 1525. The city council of Zurich had decided to suppress the small company of counterrevolutionaries who refused to have their children baptized, swear oaths and go to war.

Anabaptists now live all over the world, and the stories in this book are about one family and their relatives. My Aunt Lulu Smith kept scrapbooks which contained the stuff that tells how they really lived. Part I of *Anabaptists in America* is taken from Aunt Lulu's scrapbooks and is a bit of history of the family of C. H. and Mary (Imhoff) Smith. It is a history no more or less important than the history of countless other families who stem from the Anabaptist tree, whose lives all contain similar stories. I hope you enjoy reading and meeting some of your relatives, no matter how distant, and remembering again that we as humans are all related, not only to others of our so-called kind but to all of life…a sacred part of "That Which Is," the "Great Mystery," the "Great Spirit," the "Universal God."

> What has been is what will be
> and what has been done is what
> will be done. And there is nothing
> new under the sun.
> —*Ecclesiastes 1:9*

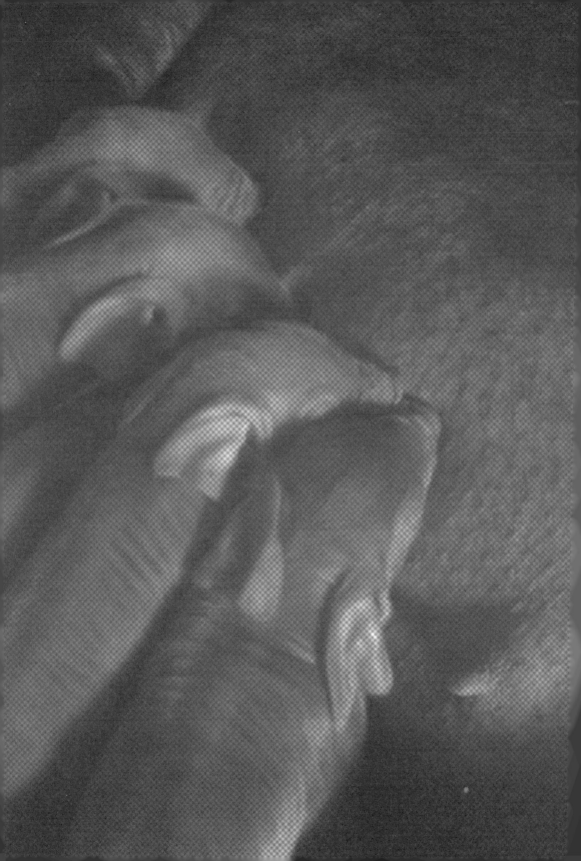

CHAPTER 2
In and Out of Academia

Chapter 2 presents Roger as college athlete and follows him through his years of work as a behavioral scientist and animal researcher before going back to the land. These articles describe how he gained academic and social validation for his academic and athletic prowess, his decision to pursue an academic career, and his subsequent transition away from the academy and back to the land that took place in the 1970s. We read about his years of studies on aggression conducting experimentation on animals, and his transformation into abandoning the laboratory in favor of animal rights. When asked in 1973 by his department chairman, "What is the most innovative thing that you have done professionally during the past year?" Roger replied, "Dear Dave, I've finally stopped torturing animals."

◀ *Feeding time for some of the recently-new-born little pigs that join us every so often at the Lake Village farm.*

Memories from my College Years

Excerpts from Get the Ball to Willie: A Tribute to All-American William Lawrence Warden, *self-published, December 2003*

Introduction

When you're soon to be 71 years old, looking back on your college days is like looking on another world. The people you knew then have disguised themselves with new looks or names, which when summed up truly fake out old brain cells that now fire to a different drummer. In 2002, to reconnect with these friends in the present, I wrote the following to my North Central College classmates:

> In a little over a year, we will be getting together to celebrate having spent a half century of great living following our 1953 graduation from North Central College. In anticipation of the celebration, I wonder if you would write a story based on your most defining memory of your North Central days. (For me, it would be the winning 1951-1952 basketball season and the start of the "Get the Ball to Willie" era.) If you would key off of this memory and then tell where your life has gone since, send it to me and I will then try to put it all together in a booklet under the title, "How We Play the Game—The North Central Class of 1953."

Once I had noted in writing the importance to me of my junior year basketball season, it came alive in my mind again. I couldn't seem to leave it alone and knew that I wanted to go ahead and tell the story. Perhaps the pursuit of trying to get a ball through a hoop at one end of a court seems trivial when viewed from a global, cosmic and eternal perspective, but it was an extremely important thing in the lives of a small group of my classmates/teammates at a given time on Earth, the early 1950s.

The story is keyed off of William (Bill, Will, Willie) Lawrence Warden who, better than anyone I ever knew personally, was able to get a basketball through a hoop. It's about who he and a few of his teammates were then, are now, and a bit of the in between. Bill Warden is probably a couple of inches shorter than years ago and he limps, not because of a sprained ankle but rather an overall sprained body. He laughs the same, has pretty much the same color of hair and is still fun to be with.

On October 19, 2001, North Central College's Sixth Reunion of Athletes of the 30s, 40s, 50s and 60s was held in Naperville, Illinois. It was time for our spirits to return to the bygone, to once again recall the feeling of that special high that comes with competitive sports, and to especially recognize football, baseball, basketball and track star William Shatzer, '42, who has been called "the greatest athlete North Central ever produced." Dr. Clarence Roberts used those words to describe Shatzer's career in his book entitled *North Central College: A Century of Liberal Education, 1960.*

My role at that 2001 reunion was to introduce the 1942-43 championship basketball team, whose 14-1 record was the best win-loss season percentage ever posted by a Cardinal basketball team. In my remarks, I noted that I too had had the honor of playing on an especially good team some 10 years later. I also noted that our team, like every other that ever ran the count at North Central, could in no way match their win-loss record. 1952 was a winning season which saw North Central match the 14 wins posted by the 40s group; we just happened to lose seven games instead of only one. There was, however, something historically special about the 1951-52 team, and that was William Lawrence Warden, the second outstanding athlete mentioned in Dr. Roberts' book. The following is what he wrote about Bill Warden's accomplishments:

> His record of 2,249 points in 79 competitive basketball games will likely become legendary to future athletes of North Central. Warden's record for total points in a season was 689, scored in 1952-

53. He received honors by membership on the NCAA Division III Little All-American Team in 1954-55. Warden's last home game on January 22, 1955 is known in athletic history as "Bill Warden Night." He was given a standing ovation when he left the game and his jersey, number 14, was retired and put in a trophy case together with a summary of his outstanding record. (Roberts, p.268)

For 50 years, Willie and I have stayed in close touch. The same is true for teammate Tom Stachnik, with whom as a fellow Ph.D. psychologist I have co-authored books and manuscripts, been a professional colleague and, like Warden, a truly close friend. Co-captain Bill Bradish I see now and then at reunions. Sometimes we talk on the phone or meet where his son is refereeing a basketball game as we sit and visit in the stands. Co-captain Don Arboe returned a self-addressed card I sent him a while back with two boxes that allowed him to check "alive" or "dead." It came back checked "alive" with a few words about his current life.

Coach Bill Olson is also a key figure in any story that relates to my most cherished memories of days at North Central. Coach passed away in the late '80s. My last visit with him was in New Orleans at the 1987 Final Four Tournament, where I met Willie and Coach for four special days at the big-time annual basketball convention. It is where hoopsters come from all over the world to gossip, check out the newest gadgets associated with the game, rub elbows with the then sacred cows of the hoop like Bobby Knight and Jud Heathcoat, and party down in whatever city it is that managed to attract the final event of March Madness.

At that time I was living and traveling in my 1971 Dodge van that I had driven over some 10,000 miles on a journey that ended with the honor of once again riding with Coach Olson by my side. The day after Indiana won the National Championship, I took him and his wife to Baton Rouge. We had an evening of dining and visiting at a hotel where we spent the final night of our long journey together before his plane departed the next day. We talked and laughed about old times, and I felt blessed to once more be at the table with this true gentleman, the mentor who affected my life in more ways than I will ever be able to tell.

Thinking back on that night brought back to mind how our 1951-52 Cardinal season would have been greatly different had Coach Olson not been the great recruiter who, with his low-key sales pitch, sold North Central College to a skinny kid named Warden.

So it is that the events of the 6th Annual Meeting of the Athletes of the '30s-'60s and the upcoming fifty-year reunion of the class of '53 inspired me to think more deeply about what my years at North Central truly meant to the rest of my life. Winning is fun and losing is sometimes hard to take. Yet both walk hand in hand until the truth of the old saying, "It's not whether we win or lose, but how we play the game," is finally integrated into the fabric of one's being. As I mentioned at the start, I am now 70 years on this Earth and still playing sports, sometimes winning, sometimes losing. I have watched many seasons come and go. All have been special in their own way. Sports have meant a great deal to me all my life and in a sense serve as the touchstone for much of my ethical perspective. But I need to get on with the story.

Warden Walks On

When I played basketball at North Central, I was a point guard. That position has always entailed, among other things, getting the ball into the hands of the best shooters. During the years in which Bill Warden was on the North Central team, getting the ball into his hands was, in a sense, the job of every person on the floor. Indeed, the team's offense, despite all-out carefully worked plays, was best defined by the words we often heard from Coach Olson, "Get the Ball to Willie."

Bill Warden appeared one evening, close to the end of the first semester of the 1950-51 school year, as Tom Stachnik, Bob Hahn and Don Neuman were haphazardly preparing for finals in North Central College's men's dorm, Johnson Hall. No one knew he was coming. No announcement. He just showed up. A mid-year graduate from Oak Park High School and member of the All Suburban All-Star Basketball Team, he was being recruited by two of North Central's conference rivals, Lake Forest and Wheaton College. On this particular day, Bill had gone to Wheaton College to talk with basketball Coach Lee Pfund and meet some of the players. He didn't know they were out of town playing a ball game, so when he didn't find Coach Pfund he drove over to Naperville to look up his high school teammate, Tom Stachnik.
"Willie!"
"Hi Tom, Donnie, Hahnie. How are things going?"
"Okay. What's with you?"
"Not a lot, just graduated. You guys like it here?"
"Yeah, it's okay, lots of fun in fact. You know how school is. You thinking of going to college?"

"Yeah, sort of."
"Where you going?"
"I don't know. Think I could get in here?"
"Sure. We did!"

Soon after, Tom took Will to meet Coach Bill Olson. In his usual reserved, gentlemanly manner and upon discovering that Bill had received offers of assistance from both Lake Forest and Wheaton, Coach was careful to not do anything that would not be in Bill's best interest.

I was a sophomore member of that 1950-51 team, and my personal first impression of Willie when he joined the squad was that it would be impossible not to like him. He laughed easily and often with a great sense of humor. More important to our priorities at the time was the fact that he shot the eyes out of a basketball hoop from a long way out, which brings up a fact that is important to note in looking at his scoring records. Bear in mind that a great portion of Bill's field goals were shot outside of what would be in today's game the three-point range. The total field goals made during his career, 847 between 1951 and 1955, translated into 1,694 points as a function of field goals. Adding in his record for most free throws made in a North Central College career, 555, we end up with the career record of 2,249, which is still tops for NCC basketball players. For those of us who played with him, our memories hold most often a scene which features Willie high in the air, both hands over his head, unleashing the ball in a fadeaway jump easily beyond the modern college three-point line, and more often than not being defended by two, sometimes three men leaping hopelessly to block his shot. In truth, no one knows how many more extra points would be added to Bill Warden's scoring record if the three-point goal had been in effect.

Between the first semester of 1951-52 and the first semester of 1954-55, basketball at North Central College was majorly defined by the scoring exploits of Bill Warden. To this day he holds every single game (52), season (689) and career (2,249) scoring record at North Central. His number 14 jersey has been retired. Beyond North Central's own campus, Bill virtually rewrote the college basketball record books, even without the three-point rule. When he finished his career with 26 points in an 89-72 victory over Elmhurst College, he was leading the nation's small college players with his 34.7 points-per-game average. Certainly there were other great college players in 1954-55. According to small-college season-end statistics released by the NCAA Service Bureau, West Virginia Tech with center George Swyers became the first team ever to break the 100-points-a-game barrier, and Bob

Hopkins of Grambling swept to a new career scoring record. Hopkins, who scored 1,036 points in 32 games, lost out to Bill Warden for the individual scoring leadership even while boosting his three-year total to 2,894 points. While Hopkins surpassed all previous career scorers, Warden finished with a better average, 34.7. George Swyers was next in average per game scoring—his 676 points in 20 games gave him a 33.8 average. Hopkins' average was 32.4.

Bill Warden was, at the time, the first player in college basketball history to average better than 30 points for each of the three seasons, even as Hopkins was the first to score more than 1,000 points in each of two campaigns. Some other stars on the court were Jim O'Hara of Santa Barbara in field goal accuracy (65.4), Pete Kovacs of Monmouth College in free throw percentage, and Tom Hart of Middlebury College in free rebound effectiveness (29.5 recoveries a game).

To gain a better feeling of the happenings during those years at North Central requires a background check that begins with the winning season of 1951-52, reading the words of those who wrote the stories and hearing the voices of those who played the games.

Roger's Story

In 1949 I graduated from Eureka High School with pretty much the same group that had started out together in the first grade. By the time I reached high school, I sort of knew I would go on to college, mainly because I wanted to play sports at that level. My family were Mennonites and there was great pressure to go to Goshen College, a Mennonite school. An aunt who had gone there even offered to pay my way for all four years. No dice. They didn't play intercollegiate sports, so my Dad and I conspired together to find and get me enrolled at a school that did. My father was an Amish Mennonite farmer who, like some of his brothers, was more interested in work, family and sports than church. It wasn't the first time he would help me skirt some of our religious traditions. While I was still in high school, he let me sign with a local semi-pro baseball team whose games were played on Sunday—a sin for Mennonites.

We checked out several schools. Eureka College coaches talked to me about going there, but it was too close to home. Illinois Wesleyan was a possibility. Bradley University was not; I didn't think I could make the team. How I ever got to North Central I will never really know for sure. I somehow heard about North Central from a distant cousin who went there from nearby Washington. I wrote for information and the next thing I knew, I was there.

It was Christmas 1945 that I was given a genuine leather, made-in-the-U.S.A., five-year diary. Written in the inside cover of this book was: "Belongs to Roger Ulrich. December 25, 1945, the Gift of Grandma Ulrich. Trust in the Lord with all thine heart. In all thy ways acknowledge Him and He shall direct thy path." I took the diary with me to North Central and everywhere else I went, including a trip to Europe in 1950, plus numerous states, and wrote in it every night for four years. The first entry was on Saturday, September 10, 1949, when I wrote: "It was my first day at Johnson Hall at North Central College, Naperville, Illinois. Slept by self, no roommates yet. Had a date Friday night, got home at one o'clock Saturday morning. It was a truly memorable date!"

Sunday, September 11, 1949: "Went to church this morning with my Big Brother, Curley Norenberg. I sat around and talked with guys coming into Johnson Hall all afternoon, then went out with Jack Will and Ron Thoresen Sunday night."

Jack Will and Ron Thoresen were two of the first people I met at North Central. After one semester at Johnson Hall, I moved to a house on Sleight Street where I roomed with Jack Will for the rest of my North Central life. He, Thoresen and I had become immediate best friends.

As I read back over the pages of Grandma Ulrich's gift, wherein each night I religiously scribbled a brief account of what I thought was important for any given day, I am more or less amazed (or bemused) at what seemed important in life for one 19-21-year-old white, somewhat observant, testosterone-poisoned male who was allegedly somewhat interested in the pursuit of knowledge. The fact is, I was simply living the truth. Years later a Native American shaman taught me that in order to learn the truth you have to live it, and the knowledge gained by sitting in classes is not as important as the wisdom that comes from living life day to day. It doesn't come easy.

Life After College

Military service was a big issue in my family, as Mennonites are pacifists in the Anabaptist tradition. Some of my uncles and cousins were in the armed services during World War II, but many more were conscientious objectors. By the time I was a possible draftee (during the Korean War), I was at North

Central. It was possible to stay out of the draft by passing a test, and many of us were glad for the opportunity. After graduation, however, I joined the Navy—in spite of all my Christian teaching to the contrary, such as turn the other cheek, love your enemies, do good to those who persecute you, return good for evil, etc. In short, I followed the national ethic which goes with being a good American. I joined up and learned how to best kill people who get in our way whenever our access to resources is threatened.

After I got out of the service (where I spent time in the Mediterranean on a tanker, played some basketball, took correspondence courses, went sightseeing and, of course, defended our country), I enrolled at Bradley University. I earned a master's degree and joined the Illinois Wesleyan faculty as an assistant dean of students and an assistant basketball and baseball coach. After two years at Wesleyan, I went to Southern Illinois University for my Ph.D. and came back to Wesleyan as head of the psychology department. In 1965, I found myself at Western Michigan University as department head and later research professor of psychology. Finally, I became emeritus professor and farmer.

Which places us in the present. I tend to get involved in issues, often regarding sports and one recently involving coaches breaking contracts. I fired off a letter to the Director of Intercollegiate Athletics at Western Michigan U. after one head coach did just that. What riles me is the trend in college sports that money means more than honoring contracts, big is better, more is better and winning is everything. On the subject of winning and losing, my years on the Cardinal five come back vividly, and in this case I wrote of a game we lost but that gave us a lot of satisfaction—the game against DePaul's Blue Devils, where five of us pretty much played the whole game against a lot of height and a deep bench. Later that year, DePaul beat the University of Illinois, a team which was ranked first at the time.

I also sounded off to our local newspaper editor about the breaking of the coach's contract, saying it seems okay for the staff to do anything if the price is right, while the student athletes must maintain grades, try to graduate and dare not take any money outside of the iron-clad rules set forth by their schools. I sure don't see sportsmanship at work here. Once again, I think of Coach Bill Olson, a true gentleman and sportsman, who taught us far more than the "fast break" strategy. North Central has never placed winning teams, money or politics as highest priority. Somehow it has been about education.

In closing, while it could be said that Bill Warden was a superstar, it is important to note that in my junior year at North Central, our team did amazing things but it wasn't only Willie who scored. In my senior year we concentrated so much on using his skills, the teamwork faltered. To make a long story very short, the lesson that struck us so clearly was the difference between teamwork—acting in community—and expecting one person to do most of the work. Not that he wasn't great at it!

Willie Warden in the middle of North Central teammates Roger Ulrich, Bill Bradish, Don Arboe and Tom Stachnik.

In Memory of Maybee Hall

June 22, 1989

Across the campus, sounds the ball
That takes away the life from Maybee Hall.
Whose ivy now prepared to die,
Prays answer please the question why?

We must progress the trustees said.
    So up goes a computer shed!
To program minds by software disks
    To worship growth, ignore the risks.
On screens that sound with imagery
    Past truths will flash their history.

See Clear Water, flow 'cross the Land.
Hear Pure Wind blow, through Forest Grand.
See Lonely Condor, in the Air.
See Extinct Species, stand and stare.

While teachers present prophecy
    That gives a false security,
And lulls young minds to live with risks,
    In scenes dreamt up by techno-gists.

A modernistic sorcery,
    That wastes away Earth's energy.
As printouts flow from software disk,
    That put tomorrow's Life at risk.

Sing, "We are the World"; set the mood,
Hear starving Children cry for Food.
Watch money addicts beg for more,
As cults of growth, deplete Earth's Store.

At the computer-versity,
    Where ethics sell so all can be,
Shot up with faith through programmed disks,
    To worship cash and shrug-off risks.
With academic sanctity,
    At saint computer-versity,
Where from the money banks we pray;
    "Oh god of state, send funds our way."
So our computer-faculty
    Can advertise on pay T.V.
Of education from off disks,
    Which daily face the virus risks.

A crack Child shot; what goes, comes 'round.
A drug firm needle soils our Ground.
"Say no to drugs," a car man pleas,
As fumes from cars drug Lakes and Trees.

At our computer-versity
    We honor ingenuity.
Midst cries for drug king-pins to die,
    We serve up booze, we even buy.
Which helps us shape Young Minds to need,
    And catch them on our hook of greed.
While trusts are sold to those who'll be
    A part of our conspiracy,
As growth sounds spread, from programmed disks,
    The siren song, that masks the risks.

So bow to me, know big is better,
    As you hunt gold beneath my letter.
Yes, I'm computer-versity
    I service all who worship me.
Material god in floppy disk,
    Whose campus popes put Life to risk.
Enroll and ply my dollar trade,
    Behave like me and get your grade.
With me, computer-versity,
    Exploit becomes reality!
From lessons taught by software disk,
    You'll learn, "take now" ignore the risk!

I sadly watched the wrecking ball,
Beat out the Life from Maybee Hall.
Now where she stood the Ground was bare,
Her soul led Home, to Lands elsewhere.
Beyond computer-versities.
Beyond the rising student fees.
Beyond new gods in software disks.
Beyond man's lies which hide the risks.

To a Spirit World, Another Place.
Where She Sings Of God's, Amazing Grace.

Now Let us Pray.
Respect the Earth.
For, How we do,
Is What We're Worth.

The Departmental Head
Changing Ethics in Academia

It's all Dave Lyon's Fault...

It was 1965 and I was happy to receive a note from Dr. David Lyon from Western Michigan University:

March 12, 1965

Dear Rog:
 At the present time our department is looking for a new Chairman. I thought you might be interested, so I submitted your name to the Dean. He will probably forward the material on the job to you in the very near future. If you have any questions and would like some "off-the-cuff" answers, I would be happy to oblige. I hope all is well...see you at MPA.

David O. Lyon, Ph.D.
Assistant Professor

 P. S. I recently obtained the description of the department that the Dean may be sending out. This description is not very good and does not really reflect the interests of the staff. For example, Mountjoy and Thor both have grant applications in, Robertson has just completed three, and I am presently working on two (conditioned suppression in pigeons and some Sidman avoidance). We have just about completed an undergraduate research proposal for the April 1 deadline. I was not going to give you a sales talk but couldn't help myself after the description.

Almost immediately after hearing from Dave Lyon, I received the following letter from Dr. Gerald Osborn, Dean of the Western Michigan School of Liberal Arts and Sciences:

March 19, 1965
Dr. Roger Ulrich
Illinois Wesleyan University
Bloomington, Illinois

Dear Dr. Ulrich:
    We are looking for a Psychology Department Chairman, starting July 1, 1965. You have been recommended as a possible candidate for the position.
    Western Michigan University is the fourth largest state-supported university in Michigan with 13,770 students enrolled. The Psychology Department is housed in L.H. Wood Hall, a new 3 1/2 million dollar building which was just recently dedicated. At the present time the staff number is 14.
    Western is growing rapidly; the estimated enrollment for 1968 is 20,000. We are doing graduate work at the master's level and hope to soon be doing doctoral work, probably around 1967 or 1968.
    If you are at all interested, your application is welcomed. If you do apply, we would appreciate receiving transcripts of your academic record and letters of recommendation. We look forward with real anticipation to your reply.

Sincerely,
Gerald Osborn
Dean

March 19, 1965
Dean Gerald Osborn
Western Michigan University
Kalamazoo, Michigan

Dear Dean Osborn:

Thank you for your letter of March 15. I am honored by your interest in me as a possible candidate for the position of Department Chairman at Western Michigan University. Perhaps the first step to be taken on my part is to indicate that I am interested. I have not found the duties of Chairman at Illinois Wesleyan University distasteful; in fact, my long-range plans have pointed toward continuing such duties, hopefully in an institution with a doctoral program.

Much of my administrative experiences have been in directing, not only academic program, but research activities as well. We have a large psychology laboratory at Illinois Wesleyan, which is supported mainly by the State of Illinois. I believe strongly in a close relation between research and application. This is not to imply that research would mainly be applied, but rather I would foster a greater knowledge of research findings among those individuals who are interested and indeed working in applied psychological areas.

No matter what else the future brings, one thing for certain is that I will wish to continue my own research. I believe that this is important for both the individual and the institution with which one is affiliated.

Without going into any further details at this time, I will simply send to you under separate cover materials related to myself of the type which I have found helpful when evaluating prospective candidates for Illinois Wesleyan. I trust that they will aid you in determining the degree to which you feel I would fulfill the criterion for Chairman of your Department of Psychology.

Sincerely,
Roger Ulrich

May 19, 1965
Dr. Roger E. Ulrich
Department of Psychology
Illinois Wesleyan University
Bloomington, Illinois

Dear Dr. Ulrich:
Western Michigan University is happy to recommend you to its Board of Trustees for a position as Professor of Psychology and head of the Department of Psychology, effective August 23, 1965. The details of your position will be arranged through Dr. Gerald Osborn, Dean of the School of Liberal Arts and Sciences.
We are glad to know that you have already indicated orally your acceptance of this position. May we have your affirmative reply in writing in the near future in order to complete our contractual arrangements?

Sincerely Yours,
James W. Miller
President

May 22, 1965
President James W. Miller
Western Michigan University
Kalamazoo, Michigan

Dear President Miller:
This is to make official my acceptance of the position of professor and head of the Department of Psychology at Western Michigan University.

Sincerely,
Roger Ulrich

June 15, 1965
Dr. Roger E. Ulrich
Department of Psychology
Illinois Wesleyan University
Bloomington, Illinois

Dear Dr. Ulrich:
It is my pleasure to be able to report to you that the Western Michigan University Board of Trustees, in its meeting on June 11, 1965, approved your appointment to the faculty here at Western. The details of your appointment will be those indicated in President James W. Miller's earlier letter to you.
We are pleased to have you join us in working to meet the challenges of higher education. We hope that your association with Western Michigan University will be a long and fruitful one.

Sincerely Yours,
John J. Pruis
Secretary, Board of Trustees

I'll be damned if it hadn't actually happened. All that talk with Tom Stachnik and Tom Mabry about moving on to a state university and how important it was to do something like that had actually come true. I was actually a living, breathing departmental head at a large state university…and really dumb as regards to what I had just gotten myself into.

The Coup

On Sunday evening, April 23, 1967, I received a call from Ron Hutchinson.

I had gone to Chicago on that weekend with my wife, Carole, and my children to talk to Scott Foresman officials about *Control of Human Behavior: Vol. II*. I had invited Dick Malott and Billy Hopkins and their wives and had hoped that, in addition to the activities that I had to engage in at Scott Foresman, we would do a night on the town. Dick Malott indicated to me that he could not go because he and his wife, Kay, had to visit his parents in Converse, Indiana. Billy Hopkins also said that he was not able to make it. I therefore went just with my family. We had stayed at the Shoreland Hotel and had driven back that afternoon.

I had gone into the lab and had returned home when the phone call was received. At that time, a number of students were living at our home, and one of them, Marilyn Arnett, along with my wife and I, were in the front room when the phone call came. Ron Hutchinson indicated to me that he, Doug Anger and Bill Hopkins would like to come over and talk to me about some matters that were very, very important.

I asked Hutch if it was something that I should have known about, and he said, "Yes, it would have been nice had you known about it."

I asked him what it was, and he said, "That's what we're coming over to talk about."

Within a few minutes, they arrived and we went into the family room, where they suggested that we close the doors so that it would be a completely private thing. They then told me that there had been a meeting that evening of the staff and Vice President Seibert, which had been called by Neil Kent.

At this meeting problems relating to my role as head were discussed: For example, I did not spend enough time in the department, I did not do things that were typically done by heads, I unfairly managed funds, I unfairly handled the assignments of teaching positions, I took the work done by Don Whaley on a proposal for an early education project and called it my own, I used students far too often to do jobs that should have been done by myself or departmental members, I mismanaged funds, I spent too much time with the Rayswift Foundation getting that particular project going and not enough with the department. In addition, I was said to be basically dishonest, manipulative, bullheaded, rigid, would not consider the opinions of others, and, most of all, could not be changed by usual techniques of persuasion.

When Hutch, Doug and Bill came to the house, my first question was, "Where is Neil?"

Ron Hutchinson replied, "He isn't here, is he?"

With those words, I was introduced to the feelings that I am sure many people have had historically when they first discovered that they had made a fantastic misjudgment in relation to those behaviors that we call trust.

For approximately seven weeks, Neil had been the leader of a group who were putting together a plan which, they predicted, would lead to my dismissal. He had arranged to meet Howard Farris in Indiana to discuss what he felt had to be done. He had then met individually with every member of the department, slowly building upon a base of discontent over events that we had originated together. The premise was that I would never submit to any major change, and it was necessary to dispose of me in a manner which involved my not knowing about the discontent in the department. It was felt that, if I knew about it, I would suppress it, the department would fold and many good individuals would leave.

In order to accomplish the coup, it was necessary for Neil to generate enthusiasm not only among new members in the department, but among the Old Guard as well. It, of course, would be untrue if I were to suggest that there were not difficulties between the Old Guard and me. I never for an instant, however, had mistrusted Whaley, Hopkins and Malott, who, along with Kent, led the coup.

Among the other members of the department there was much concern that the whole thing was wrong. This held especially true for Doug Anger and Ron Hutchinson and only to a slightly lesser degree with Rob Hawkins, Dave Lyon and Paul Mountjoy. Both Doug and Ron were individuals with whom I had been acquainted for a long time. The tactic used to get them to Billy Hopkins's home that evening and to involve them in the coup is worth noting. It had to be done in such a way that they did not know what was up. Their attendance was accomplished in this manner (in case anyone else is ever interested in university politics and discovering how you move to get rid of a department head):

When everybody had known what was up and had sworn to secrecy except Doug Anger and Ron Hutchinson (who were close friends), it was decided that Doug Anger would be the next person to be brought into the inner circle and told what was in the air. In order to do this, Doug was called into an office with Neil and several others and made to promise that he would not divulge the information that he was about to hear. He was told that it was critical to the survival of the department. He was then advised that it could not be passed on to him unless he swore that he would not pass it on. Here a prediction was made of the nature of Doug Anger by a wise student of behavior—Neil Kent. Neil knew that if Doug promised that he would not tell someone, *Doug Anger would not tell*, regardless of how he felt about the information.

Upon being told what was up, Doug immediately began to gather evidence for my support. He carefully took notes at every meeting and set out to find data which would either substantiate or disallow the charges that were made against me. Doug Anger became my defense attorney, and Ron Hutchinson subsequently became the manager of the counterstrategy.

Ron was not told anything about what was happening and only learned about it the evening he went to Bill Hopkins's home, where he reported being so angry that he could hardly talk.

Some felt that it would not be bad for me to be released from the headship but felt that the method being used was unethical. Malott staunchly supported my leaving the university. Paul Mountjoy and David Lyon were staunch defenders of my continuation in the department and basically confused as to whether or not it would be a good thing for me to leave as head. Malott and Whaley were of the opinion that I should be fired.

After leading the whole affair, Kent adopted a more neutral role. He felt that it was probably a good idea for me to no longer continue as head. His view on whether or not I stayed in the department depended a lot on what I wanted to do and what the department felt was best. Neil was a master at deciding what he wanted and then getting others to go along with it. Always he did it in a way so that it looked like he was only doing the bidding of others.

The charges made by individuals were largely contraindicated by the data carefully researched and gathered by Doug Anger. This proved to be a deterrent to the efforts of those pressing for my actual dismissal. Ron Hutchinson was finally able to convince Dick Malott that he could trust him. The plan was then laid before him: The strategy originally decided upon was to have Neil Kent take over as head of the department. This revelation, however, came somewhat later.

At my home that evening, I, in the manner that certainly could be predicted and was predicted, was fit to be tied. I had a long history of countering aversive stimulation with counter-aggression. I had had a long history of not taking things sitting down and being perfectly willing to fight back and fight back hard. It was suggested to me, time after time, that evening to just sit patiently and listen as opposed to always talking.

With the various charges and demands written down, Doug and Billy left. I had given them my agreement to meet the next day with the whole department and let them know what my plans would be. It was 2:30 in the morning. I had from then until late morning to make a decision. A meeting was called for approximately 11:00 a.m.

Ron remained behind after Doug and Billy left and said, "All right now, I'll stay. Let's get it together and decide what we're going to do." He indicated the unequivocal support of himself and Doug. Out of this experience, I learned not only the lesson that trusted individuals can default on your expectations of their trust, but, more importantly, I learned about unfaltering devotion and loyalty that friends can show to other friends when they feel that such support is needed.

In this regard, I will never be able to forget the kindness and the sincere efforts of Ronald Hutchinson and Douglas Anger, who both showed tenacious support. In the same vein, the extent to which my students were willing to go along with whatever I wanted was another lesson to me, which perhaps as much as anything else helped change some of the directions that my lifestyle was to go.

Two days after the Sunday meeting, George Hunt and Marshall Wolfe stepped into my office saying that they had heard rumors that I was in some trouble and wanted to know if there was anything that they could do. Both indicated that there would be no limits as to what they would be willing to do. Overall, the major role I needed from all my friends was that of modeling behavior which falls under the heading of "keeping it cool."

In terms of understanding me and predicting my behavior, Neil Kent and others did a good job. For example, it was pledged by all members of the department that no one would meet with me individually. This, of course, was a pledge made by neither Doug Anger nor Ron Hutchinson. Prior to going to the staff meeting the next day, I met with Dean Seibert, Dr. Osborn and Dean Lowe, all of whom were taken aback by what they described as a first in Western Michigan University's history, that of an actual coup being attempted. Given the structure of Western Michigan University administration, it was quite clear to them, they said, that this could not be allowed to happen and that I had their total support.

Dr. Russell Seibert: The Alternatives

Dr. Seibert wrote on a strip of paper that I should go back to the department and give them the following alternatives: Letting them know that I wanted to

(a) continue the headship for two more years, or that

(b) I wanted to continue in the headship for an indefinite period of time.

In both instances, I was to let them know that I had the university's backing on this issue.

"Roger," said Dr. Seibert, "I know that you are a very persuasive individual and you are very proud. I know that this is an unhappy set of circumstances for you. I also know, however, that you'll work it out all right. Do the best you can and know that you have our support."

At that point, Dr. Seibert tossed me the ball and simply indicated that he would back me on whatever it was I felt that I must do.

What I felt I must do, I must admit, fluctuated sometimes from second to second. Even in the midst of stressful situations, there was much mirth. There were many instances in which Ron Hutchinson, Doug Anger and I could step back and laugh at the whole situation. Hutch described it one day:

He said, "You know, I guess it's not too strange when one looks around and sees what's happened and when you consider what you've done in the past couple of years. It's sort of like taking a whole bunch of people, none of whom were ever acquainted with the behavior research at Anna State Hospital, and stuffing them into a tin-roof house where they were to remain and upon which you beat until they demonstrated that they could work hard."

In my laboratory, I was studying the effects of aversive control and was well acquainted with what long hard ratio runs and stressful situations do in the area of aggression. So it was of little consequence that staff members' wives frequently found their husbands now at meetings or at their laboratories at the university. Only a few short years before they had been able to break away at 3:00 in the afternoon and work on the patio. I was not sufficiently acquainted with how to use leisure time and still be productive, to model that approach or even tolerate it when I first came up to Western Michigan University. The fact that the "New Guard" had behavior patterns very similar to mine did nothing to lessen the problem.

Dick Malott was in many ways very similar to me in his lifestyle and the way he looked at things. In his case, the situation had the effect of making him even more paranoid regarding his trust in the direction I could go. His lack of trust in people generally and the extent to which he was well acquainted with his own strategies made him doubly suspicious of me.

After the meeting with Dean Seibert, Dr. Lowe and Dr. Osborn, I went over to the student center where the meeting was called. Already gathered there were the majority of the members of the department. Ron Hutchinson announced in a loud voice, "For goodness sakes, don't sit by me." I smiled and sat next to Dick Malott and asked him how things were with his parents. He mumbled, "I didn't make it down there, man." I acknowledged that I understood that he had not and then suggested that we must keep up the fight, however, to get rid of dishonest people.

With Dick Malott and with certain other members of the department, there had been a basic disagreement between them and myself for some time. Although they considered themselves to be hard-nosed behaviorists, I often looked upon them as being quite unrealistic about how behavior really worked. I felt that they were continuing to look at things, far too often, as being either black or white, good or bad. Rather than following their own supposed teachings that behavior does have reasons and dealing with it in a tolerant way, they tended to blame the organism as if it could be something else.

They, like others, talked about the use of positive reinforcement and supposedly agreed that organisms do what they can to increase the probability of pleasure and decrease the possibility of pain. In the next breath, they got angry when it looked as if someone was doing something for no other reason than to increase the future probability of his receiving pleasure. Time after time, I would argue that that is just the way things are and that we all behave in that way. One evening Don Whaley said to me, "Roger, you tend to do things only if there is a chance that *you* will get something out of it."

"Yes," I said, "I think you're right."

It was very obvious that my behavior patterns had hurt a lot of people, and it was a reasonable expectation that a change in my actions would take place. For example, I was learning that I must lower the frequency with which I was willing to follow responses made by others with immediate feedback of an aversive nature. Such actions on my part had the tendency to suppress feedback in a rather dramatic way. I was extremely impatient, and the extent to which I could barely tolerate slow, easy gain was very much a part of the problem.

Another thing that was very revealing to me was how much I really hurt as a function of the activities that surrounded the loss of support by Neil Kent. I am afraid that for some months I was bitter about those events and often behaved in an immature and pouting fashion. Once again, my own verbal behavior did not marry with my actions. I was not tolerant of other individuals although I could verbalize knowing that they did what they had to do.

My emotions had not yet caught up with my verbalization and my knowledge. I was not able to remain compassionate in the face of extreme pain I felt from actions of people who were dear. Also, I was concerned that I could indeed have behaved in a way that would make someone who had been as close to me and as trusted by me as Neil Kent end up working as diligently as he did to get rid of me. I simply had to start re-evaluating my behavior patterns.

I met with the department that morning and indicated that I was not able to give them a decision at that time. I went on to say that, as far as I knew, it looked reasonable for me to eventually consider leaving the position of head.

At that meeting, David Lyon made a strong plea for the good things that I had done in the department. He said that he personally was not about to support my discontinuation as a staff member. He did feel, however, that the department would be better off if I were to continue in some capacity other than head. His point was that the headship was unnecessarily restricting me.

David, along with Mountjoy and a few others, never made me feel that they were blaming me. Jack Asher, who was at this meeting, was also not unkind. In fact, he simply indicated that I was not willing to take "no" for an answer. And with this I could not disagree, particularly as related to Jack, as he had given me more practice in this regard than almost any other human being that I had ever met. Various things were discussed and, again, no general conclusion was drawn.

I left the meeting with the intention of trying to figure out ways in which, if nothing else, the whole program could be held together. It had become the case that I had identified so closely with the department and the headship that it was difficult for me to separate out what was good for me, what was good for the department, what was good for the profession, what was good for others, etc.

Many of the criticisms that were leveled were simply criticisms of the way that life really was and is. "Roger Ulrich does everything to enhance his own position," they said. That's true. I was willing to admit this. The only problem was that admitting it got me even more problems.

"Roger Ulrich will work harder than almost anybody I've ever known," they said, "to figure out ways to get rid of things he doesn't like." That was also true, but, once again, the difficulty in admitting it often caused more problems.

I was finding that honesty and candor carry with them their own special kind of problems. Often when one is not skillful at hiding his true feelings, he is going to have problems. To an extent, I, as a head of a department and full professor, was willing to admit that heads of departments are like everyone else, dishonest at times with deans and vice presidents as they try to get more money. Politics on the grand scale or just in everyday life requires, if one is to be successful, a very careful assessment of the outcome of one's actions. Oftentimes, this requires that one carefully monitor his actions and thus keep hidden from other individuals the truth of his feelings.

Learning to Lie with Class

To be too honest about the way we really are, the way we really feel, can prove to be one of the most devastating things that one can do to another individual. More than anything else, it simply reminds them of how basically dishonest they themselves are. All institutions have certain rules by which they play the game. They, via their administration, go out and seek money. In every case, the rituals are pretty well put down. To the extent that the dance is danced not too much differently than the way it has been danced before, it will not upset people. To deny that each individual, as he does his own dance, often does not report facts exactly the way that they have occurred, or the way they might be predicted to occur, would be foolish.

The trick is to practice dishonesty in such a way that it does not upset people. Lie with class. This is true with the institution of marriage; this is true with the institution of friendship; this is true of all mankind. As you move toward your most cherished goals, you behave in a way so as to not hurt others, but rather make them feel as good as possible. It goes without saying that many, many times were you to let others know certain facts, you would definitely hurt them. When you hurt them, another person might try to stop the hurt by punishing you for what you did to hurt them. Also, they might try to escape or avoid you or reward you for other things you do that don't hurt.

In one sense, I was not as adept at subterfuge as is necessary to be a good department head. Only in later years, after becoming a research professor, did I begin to become more adept at this particular area. My earlier approach was simply to avoid those situations which required me to behave in a way which was basically contrary to what I felt and to what I wanted to do.

Change takes place and there is no denying it. Change is also painful, and it seems that the more rapid the change the more painful the hurt. Quite frequently it hurts when one is forced into a different way of existence. The change of growing old hurts. Knees that once bent very well and legs that did not tire when riding a bicycle or when running now start hurting. Aging is painful. The same may be true of institutions. As they age and as others come along and cause them to try and change, the composite behavior of institutionalized man begins to hurt. When you, as an individual, can be identified as an agent of change, you will become associated with the hurt that goes with change, and there is no way to escape it.

To the extent that I was identified at Western Michigan University as an agent of change, both in the Department of Psychology and throughout the university, I caused people to hurt. I was recognized as a change agent who was willing to move quickly and willing to produce rapid change. The fears that surrounded the amount of pain that I produced were greatly felt, and the fears that I would continue to move in this direction increased. Living institutions and the men (*sexist 60s...*) that populate them tend to protect their composite self in much the same way that an individual protects himself. The excitement of change is great. To dive to the bottom of the ocean, to climb a mountain, to speed on a fast-moving motorcycle or race car is exciting, but they all carry with them some danger of pain.

Research Professor: There Is no Searching outside of a Real Situation

At Western Michigan University between 1965 and 1967, the Department of Psychology changed a great deal. The pleasure and exhilaration of being a part of that change tasted like the most excellent of wines. We drank deep only to experience an after-the-party drunk. How to be an agent of change without producing pain for one's self and others is a difficult question. I eventually decided that, although I had the support from the administration, the expectations that the majority of the Psychology Department members had for me as head were not the same as I had for myself. Thus, based on all the inputs available to me, I made the response of resigning as head before the 1967-68 academic year and assumed the position of research professor in the Department of Psychology at Western.

June 29, 1967
Dr. Roger E. Ulrich
Department of Psychology
Western Michigan University

Dear Dr. Ulrich:
Western Michigan University is happy to recommend you to its Board of Trustees for a position as research professor in the Department of Psychology, effective August 21, 1967, at a salary of $15,000 for the regular academic year of two semesters. The details of your position will be arranged through Dr. Paul Mountjoy, Acting Chairman of the Department of Psychology.

As research professor you will be expected to devote one-half time to teaching and one-half of your time to research unless other arrangements are made at some time in the future with the approval of your department chairman and dean.

This appointment is meant to recognize your contributions in the field of psychological research and give you assurance of an opportunity to continue your activities in that field.

We hope that you will find this offer an attractive one and that we may have your affirmative reply in writing in the near future.

Sincerely yours,
James W. Miller
President

As I have mentioned before, Doug Anger, Ron Hutchinson, Paul Mountjoy, David Lyon and other colleagues impressed me with their behavior in a way that will have my eternal gratitude. The behavior of Neil Kent, Bill Hopkins, Don Whaley and Dick Malott must be categorized, in the end, in exactly the same way. Both groups did what they had to do at the time and the outcome was fantastically beneficial to my own future development and that of the Department of Psychology.

The events which surrounded my changing from head to research professor were both painful and gratifying. No doubt we all learned, and, hopefully, the lessons which surrounded those events will carry a meaning for all of us involved for the rest of our lives, and perhaps for others as well. Men and women do what they have to do. Hopefully, we can learn our lessons from what occurs in the past and pass them on so that those who follow will be even more effective as agents of change. I believe that we must expect failure to bring even more rapid change.

As individuals and their institutions move toward the future, they have to learn to stand the pain of change perhaps better than ever before in history. Our technological world is changing, and it often appears that human change is not keeping up the pace. We must, therefore, begin to vigorously look for ways to better educate future generations to tolerate change and move with it before our own destruction hurries upon us so quickly that we lose in our race for continued existence sooner than might otherwise be expected.

Toward a More Perfect Union

Introduction

What follows in this report is simply a record of some of the words bandied about between October 5, 1977, and April 8, 1981, which, for me, related to the issue of academic freedom and tenure. It is now 1996. I am still here as a colleague to all the WMU family. I feel we should all be thankful for the privileges afforded us by academic freedom and tenure.

Conscientious Objector*

In 1976, after more than a year of watching the flow of words which passed between Western Michigan University (WMU) officials of the WMU branch of the American Association of University Professors (AAUP), I was convinced that the point of contention more than anything else was a desire for more money. Regarding the issue of whether or not the local AAUP chapter should become a bargaining agent for the faculty, *I voted "no."*

Prior to the time of the vote, I had been a member of the AAUP because I believed that it was the group most concerned with guarding my freedom to do research and publish my results regardless of what they may be and how popular they were with current social thinking. But at the time of the vote, it seemed the AAUP was standing not so much for my freedom as a behavioral scientist, but for something for which I had less interest: more money for a

---

* This chapter is excerpted from an October 5, 1976, letter I wrote to "Colleagues of the AAUP."

group who, relatively speaking, was already well off during a time in history when many of our friends and neighbors had less than enough, especially friends and neighbors in the so-called Third World countries.

Given what I read coming from the leadership of the local AAUP chapter, I became convinced that its interests were no longer in accord with mine and thus ceased my association with it. Letters urging me to join kept coming, and I became angry when I heard the AAUP was intending to negotiate an "agency shop" that would force suspensions and even dismiss tenured faculty who refused to adhere to demands of loyalty via membership dues or equivalent fees paid to the local chapter.

In one especially distasteful letter, the AAUP suggested the administration would use dismissal to vastly increase its discretionary powers, but the letter hid the fact that the AAUP would be willing to be at least as ruthless.

I thanked the AAUP for letting me know that there was now, more than ever, a need for us to be unified. Nevertheless, I added that I saw no indicators which would make me change my position in favor of the AAUP's desire to force me into joining its efforts. The only thing I saw faculty getting out of the whole mess, if I went the union way, was two administrations with which to cope. "Now I ask you, comrade colleague, what the hell is good about that?"

Another thing that came to mind was that maybe it wouldn't hurt faculty members and administrators alike to have a little less. Our homes are too big, our cars are too big, and far too often so are our bellies and mouths. I know it's tough to kick the habit of being money junkies, but it wouldn't hurt for us to cut back a little.

One time in St. Louis I parked an old beat-up car of mine on a back street to watch a baseball game between St. Louis and New York, and a kid came up and said, "Hey mister, I'll watch your wheels for $1.00 to see that nothin' happens...." He was about nine years old, black, and he had class, so I paid. I didn't feel that the AAUP had class. I wasn't even sure it was honest or that I could trust it to "watch my wheels."

Another thing that concerned me was that I heard my neighbors saying that they didn't want as much of their tax money going to pay professors. They weren't sure we are worth the money we are being paid. Most people who are pros know that now and then you have to chance a salary cut; maybe you even get hurt and can't play any more, like Jimmy Hoffa. He had class!

I felt that if we were all so dedicated to our fellow WMU colleagues, instead of taking a raise we ought to take a cut and give it to our less fortunate colleagues, so they could stay and teach and do research and not get canned.

In any event, we might ask faculty and administrators alike if it is possible that we are spending too much time getting rich and not enough doing research and excellently teaching.

Personally, I'd still rather pay a poor kid the equivalent of my alleged union dues to watch my car than give the union the responsibility of making better use of that amount of money.

In October of 1976, I received a WMU-AAUP newsletter prepared by a man who also sent me notes to get me to vote for a legislative candidate whom he called a liberal. And so I wondered if liberal meant being willing to fire faculty because they refuse to pledge allegiance to the local AAUP. Apparently so, for the editor informed me that in order to ensure financial security to the faculty's bargaining agent, so as to provide as great a degree of fairness as possible, I must join and pay dues to a group which I was coming to perceive as unethical. Or if I chose not to join, pay the service fees assessed on nonmembers:

> "As a condition of continued employment, all persons in the unit shall, by November 1 of each contract year or within thirty (30) days after the effective date of appointment, whichever shall occur later, tender annual payment to the Chapter of either the dues assessed on members, or the service fees assessed on non-members or sign an authorized form for payroll deduction of dues or service fees.
>
> "Following at least fourteen (14) days written notice to the faculty member, the Chapter may request the termination of a faculty member who has not complied with the dues or service fees section of the Article. The Chapter's request to Western shall be in writing. Upon receipt of a proper request, Western shall terminate the appointment of such faculty member as of the end of the semester in which the Chapter notifies Western that said faculty member has not paid Chapter dues or service fees."

If I refused to be intimidated by a group of men and women who wished to force their views on me, I was now in danger of being dismissed by Christmas. So I thought, "Is it indeed the case that the WMU chapter of the AAUP is saying that the tenure right—which has for years allowed faculty to speak out on their findings and beliefs without fear of oppressive retaliation—is now being removed and that the AAUP wishes to become the very agent who is demanding my termination as research professor at WMU? Should I refuse to engage in an action to which I am morally opposed? Has the AAUP become the witch hunter who says 'sign our loyalty oath or else'?"

I was certain that the AAUP position was morally and ethically indefensible. But I also thought that the question of it being legally defensible was perhaps a different matter. Legal defense often depends on how much money you can raise, which was, no doubt, one of the reasons for the attempted monetary shakedown in the first place.

The AAUP's guilt over the issue is undoubtedly what may have prompted their citing some words by an alleged fact finder, Judge George Bowles. The September 22, 1976, WMU-AAUP newsletter stated that the judge's comments on, and especially his reasoning about, agency shops deserve thoughtful study or review because they go to the philosophical, legal and political heart of the matter. That's true, but not for the reason the AAUP suggested. The AAUP, of course, wanted its position to look noble. Hell, who doesn't! So did Hitler and Nazi Germany, and all other individuals and groups who are trying to steal something of value without the victims noticing or fighting back! But what to me was a rationalization continued, "We, the AAUP and Judge Bowles, find no philosophical difficulties with the proposition that all should pay a fair share. Those who reap the benefits along with active members should pay their fair share." What benefits? Benefits according to whom? "Do I still have the freedom to disagree with Judge Bowles," I asked, "or does the AAUP feel that that has been already taken from me? *If they do, they are wrong. My agreement is not theirs to give or take. That's what academic freedom was and still is all about.*"

Furthermore, why should I believe Judge Bowles to be a fair man when he says that the association proposal does not do violence to established systems to underwrite the cost of a program calculated to benefit all within a given group when he uses as his example that everyone, or almost everyone, pays state and federal income taxes? No doubt he added "almost" since he would, of course, be aware that many of the rich pay little if any state and/or federal income taxes. Or maybe he just read *Bury My Heart at Wounded Knee* or has talked to some American Indians, Chicanos, poor blacks and whites and other minorities who are not convinced that they have indeed reaped the benefits of programs designed for their alleged welfare.

It seemed to me that the WMU-AAUP chapter was simply doing something that has often been done before—trying to pull off a little legal robbery. And the other bandits, the administration and the Board of Directors at Western Michigan University, agreed to the crime.

When I was about 10 years old the Second World War came, and I, being the child of Amish Mennonites, began to learn what it meant to be a conscientious objector. To some it meant you and your loved ones were yellow cowards afraid or unwilling to fight. To others it meant draft dodgers who wanted to stay at home and get rich while others got shot, or yellow paint and rocks thrown at your church. And often it meant a courageous commitment to the ideal of nonresistance that Amish Mennonites and other Anabaptists took with them from one country to another as they searched for a place that would finally allow them freedom of conscience, a freedom which often meant facing death and severe persecution to show how much you really wanted it.

"Although I disagree very strongly with many aspects of what I perceive the AAUP to stand for," I wrote to the union leadership, "I feel confident that we are still among colleagues who will respect one another's right to dissent. It is in this spirit that I affirm my conscientious objection to the AAUP's stand in regard to the consequences for not joining or paying their fees. Also, let it be understood that this is not a request to the chapter for 'status' as a conscientious objector. I already have that status. I am Roger Ulrich, tenured research professor at Western Michigan University, a strong objector to what I perceive to be a stand on the part of the AAUP which, in essence, seems to say 'disagree with us and we will purge you from our midst.' Please understand that I am not oblivious to the fact that the administration also signed that contract. So I'm not looking for their help in my corner. And I'm not concerned about the dues and fees."

I closed the letter diplomatically, stating that "I can't join you now, maybe never, but I will try in the future to understand your side and would be open to receiving information from your group that might clear up some of the points raised in this letter."

March 13, 1979
Dear Professor Ulrich:

This is in belated response to the letter you sent me several months ago in which you enclosed a letter dated November 20, 1978, which you wrote in order to inform various people of some developments at Western Michigan University regarding the problem of conscientious objection to membership in the faculty union.

I was much involved in the same kind of problem during my years at the New York Civil Liberties Union, particularly involving members of the United Federation of Teachers, the union that represented public school teachers below the college level in New York City. I am very sympathetic to the views you express.

In general, ACLU policy reflects the concerns of the labor movement of many years ago. I am one of those who believes that we should reexamine our policy with respect to the problems you describe in your letter.

As you may not know, even in union contracts that do not contain a conscientious objector clause, a dissident employee who does not choose to become a member of the union and pay dues, does not have to pay a service fee equivalent to dues. Though a service fee must be paid, you may deduct that amount which is attributable to political activities of the union to which you object and you are required to pay only that amount which is attributable to the collective bargaining service that the union provides.

Of course, it is almost impossible to exercise that right since the union is not obligated to provide you with statistics on its expenditures so that you can determine what proportion is going for political activities and what proportion for collective bargaining. Moreover, I understand that your objections go directly to the content of the contract itself and as such would not be covered by the right I have described.

I would appreciate it if you kept me informed of your continuing struggle at Western Michigan. I am interested in the issue and would like at some point to raise it with the appropriate committee of the ACLU.

Sincerely,
Ira Glasser
Executive Director
American Civil Liberties Union

2 April 1979
Don Lick, President
AAUP-WMU Chapter

Dear Don,

We understand that the WMU Chapter has asked the dismissal of Roger Ulrich, a tenured professor. If this attack on tenure is successful, we want you to know that we might well decide that we can no longer conscientiously support an organization that strikes at a principle we feel is essential to academic freedom in the university.

>   In the past we have supported the fight of persons wrongly dismissed by the previous administration of Vice-President Mitchell, and indeed one of us experienced a long grievance struggle with the university over the granting of tenure. We therefore have every reason to feel that tenure as a principle is far more important than collective bargaining, about which we are beginning to have very serious doubts. Hence, while we have little sympathy for Professor Ulrich's intellectual position in particular or for behaviorism in general, we defend his right to remain at this university.
>
>   We remember from bitter personal experience how easy it was to be fired for one's beliefs during the 1950's, and feel that the AAUP had better consider very carefully what pressing the Ulrich case might mean.
>
> Sincerely yours,
> Clifford Davidson
> Audrey Davidson
> cc: Roger Ulrich

The Freedom to Work

The following American Civil Liberties Union (ACLU) policy statement regarding "closed" union shops was sent to me by ACLU legal director Bruce J. Ennis. My response follows.*

Policy #46

Closed Shop, Union Shop and Right to Work Laws

(a) Requiring an individual worker to contribute financially **to the labor union,** which acts as his or her statutory representative, an amount to equal to his or her share of the costs of negotiating and administering collective agreements does not violate the individual worker's civil liberties. **Because the union has both the authority and the responsibility to represent all employees in the bargaining unit** and because the collective agreement negotiated by the union regulates terms and conditions of employment, **the union can be properly said to be an instrument of the worker's industrial government.** In this respect, there is a substantial and proper analogy between labor unions and government. **It is generally accepted that a citizen who receives the benefits of democratic government should share in its costs; similarly a citizen of industrial government cannot properly claim a right not to contribute to its**

---

* Emphasis added by author.

cost. The foregoing essentially comports with provisions of existing federal labor statutes which permit labor-management contracts to require individual workers, as a condition of employment, to tender regular union dues and initiation fees to their collective bargaining agents.

Requiring an individual worker, as a condition of employment, to do more than contribute to the collective bargaining agent his or her share of the costs of representation, by assuming the other obligations of union membership or to become a union member, is a violation of the worker's freedom of association. Compulsory membership that requires an individual to become part of an organization with which he or she fundamentally disagrees, or which he or she does not choose to support, requires the worker to be bound by its rules and decisions beyond those made through the negotiation and administration of the collective agreement. **The principle of freedom of association includes the freedom not to associate.**

Inclusion in collective agreements of union security clauses, which by their wording appear to require employees to do more than contribute financially to the union, is an encroachment on the individual employee's freedom of association, for it misleads the employee as to the basic rights to which he or she is entitled.

Union constitutions should contain explicit and reasonable rules for resignation from the union. Failure to include such rules misleads members as to their rights and obstructs their exercise of those rights.

Requiring individual workers who have chosen not to be members of a union to support financially the political activities of the union is a violation of the individual's political freedom.

Requiring individual workers to contribute financially through labor unions to political candidates' campaigns to which they do not wish to contribute is a violation of the individual's political freedom. This applies both to members and to those who have chosen not to be members. Union funds should be allocated or assessed for political campaigns only in ways that provide reasonable opportunities for workers to certify that they object to such expenditures and not to contribute to them.

Requiring an individual worker, as a condition of employment, to become a union member or assume the obligations of union membership *before* hiring (the closed shop) is a violation of freedom of association.

There need not be a legal obligation to make special provision for those who demonstrate conscientious objections against support of collective bargaining representatives. But civil liberties are better served if employers and unions do respect such claims of conscience. [Board Minutes, February 17-18, 1973.]

(b) The ACLU emphasizes that it does not presume to evaluate the merits of union policies concerning economic matters, to prescribe the ideal governmental structure and procedure for a union, or to decide which union should represent certain employees, any more than it would presume to decide the policy, method, or structure of the United States government. It maintains a strict neutrality with regard to unions, as with regard to government.

It is not without significance, however, that the history of interference with the civil liberties of labor organizations in the areas where most right-to-work laws have been enacted gives ground for concern that such laws carry the potential danger of being used—or misused—directly to obstruct the exercise of basic organizing rights.

Nor does the ACLU's interest in protecting labor's right to organize, as an expression of freedom of association, carry any implication of a civil liberties requirement that labor's organizing efforts must always be successful. [Board Minutes, January 3, 1955, May 28, 1956; Minutes of Labor Committee, February 9, 1959; Staff Statement on the ACLU and Compulsory Union Membership, 1965; News Release, February 17, 1955.]

POLICY UNDER REVIEW, along with other labor policies.

March 19, 1979

Dear Mr. Ennis:

I have read ACLU Policy #46 very carefully and I think I can understand why someone not familiar with my situation at Western Michigan University might feel, based on that document, that the ACLU might not be able to give me support. I note, however, at the end of the policy statement there is written "Policy Under Review." I find this statement gratifying, for I am convinced that such a review is needed. Indeed, it is my conviction that there will never be written a policy that protects one's civil liberties against each and every potential new affront.

The attempt on the part of the leaders of the WMU-AAUP faculty union to coerce me into actions that I cannot, in good conscience, perform is a very serious threat to the integrity of the general concept of academic freedom. More serious, however, is the potential danger to my rights as described in the first amendment to the Constitution of the United States. I am a psychologist, a behavioral scientist and a research professor. I was appointed as such by the Board of Trustees of WMU and given the mandate to explore the realm of human behavior and to report the results of this exploration in my classroom as a teacher and around the world both as a speaker and via my publications. To me, this mission is as sacred as any that a human being can have, and to me it is my religion.

I feel that it is extremely unfortunate that the leaders of the faculty union movement, who consider themselves analogous to labor, have come to have so much in common with a university administration who consider themselves analogous to management. As such, both have seemed to lose sight of the essence of university research and teaching. For me, teaching and especially research is a profession which requires human actions that are not specified by any union contract. No person can write a contract telling the explorer where or how to make a discovery, nor can they specify what the researcher must say as he or she attempts to teach others the results. As Leon Knight, another college professor who has similar concerns, says, "the idea of academic freedom, the idea of the dissident person, the idea of a person who marches to a different drum, is very precious. And yet unionism is coming in and saying I must march to that drum. If they can determine not what I teach in the classroom but whether I teach at all, this is the ultimate threat to academic freedom". I, of course, would add that in the case of the teaching and research profession, you cannot dismiss a person from a university position which requires that he speak out as a teacher both through the spoken and written word without compromising that person's rights as defined in the first amendment.

Perhaps it would be well at this time to review the circumstances which have led to the point of this letter. The following point is very important: it is not the case that my sympathies are not with the faculty! Generally speaking, the role of the faculty was and is far too subservient to that of an administration whose job it was originally to augment the furtherance of teaching and research and, above all, student welfare, but has come instead to be generally concerned with perpetuating itself. It was this issue that I personally felt and continue to feel should be addressed by the faculty leadership.

Because I felt the union organizers were not effectively addressing the proper issues, I was not especially supportive of their efforts. Furthermore, as a researcher I was allowed a great deal of freedom and received generally good support at Western Michigan University.

It was at about this time that the bigger issue arose. I discovered that the faculty union leadership was proposing to insist that all faculty be members of the union or pay service fees, or be dismissed from the university. Feeling strongly that **the principle of freedom of association** includes the freedom not to associate, *I wrote a detailed letter to the faculty union president expressing my conscientious objection.*

Perhaps because they were just getting started, or perhaps because they felt that civil liberties are better served if employees and unions do respect such claims of conscience, *they chose not to require that I pay the service fees. Because mine was not a concern solely for money but rather a concern for the loss of the freedom to dissent, I agreed to a compromise which allowed for sending an amount of money equal to the union dues and/or service fees to an organization of my choice. This also was agreeable to the union.*

In so allowing, the union established the fact that they didn't need that share of the costs for negotiating and administering the collective agreement with the WMU administration. Furthermore, as is stated in the ACLU Policy #46, the union has both the authority and the responsibility to represent all employees in the bargaining unit. As long as they give me the option to dissent without the threat of dismissal, they do indeed represent my sentiments as a member of the bargaining unit.

Now the ACLU states that it is generally accepted that a citizen who receives the benefits of a democratic government should share its costs. This I do. I pay local property taxes that support educational and other systems of this state. I pay state income taxes that do the same, and I pay federal income taxes that do the same. Seeing that the amount of money which was each year coming in the form of a raise did not offset the rising cost of living, I began to research the issue and discovered that it was persons directly on the payroll of my democratic government, such as members of Congress, judges, both at the national and state level, who were receiving the larger cash benefits.

I discovered it was the members of the Senate and the House of the State Legislature with their staffs, and the appointed judges and the mental health workers and the people in the State Department of Education and Social Services and the faculties and administrations of our state universities, who were the citizens who seemed to be receiving the greater benefits of this democratic government.

In short, I felt the amount of money that the WMU-AAUP union leaders were attempting to win in the form of raises and benefits was in no way equal to what would be lost in terms of abridging one's freedom of speech should the faculty union, with the consent of the university administration, be allowed to fire dissenters.

I am becoming more convinced that the analogy to the history of unions in industry and other areas is not appropriate on a university campus. The necessity for a culture to hold sacred a citadel as a potential base for dissent against practices which undermine the civil liberties of the citizens of this Earth is extremely important. That the university is the proper base for this citadel seems apparent in that it is the university which has been assigned the formal role for the transmission of cultural values. We are responsible for researching, educating and granting degrees to citizens in every known area of human concern—for example, law, medicine, business, science, the arts, religion, philosophy, government ethics, and all other areas in which we grant degrees. The model we provide both reflects and projects the culture of the past and the future.

My conflict at Western Michigan University, therefore, is not solely with the WMU chapter of the AAUP, which serves as the faculty bargaining agent, nor is it solely with the administration. More realistically, it is with a progression of events which has allowed or produced in all of us at the university a pattern of behavior which is over-consumption almost to the point of an addiction. My not joining the union nor paying dues was an attempt to speak out on this issue.

My years as an animal researcher have taught me that there are other ways to communicate. There are nonverbal forms of speech which I believe should not be abridged, just as the first amendment talks of non-abridgement of verbal speech. It is this type of expression that I contend is presently being threatened by the contract between the faculty union and the administration.

How can one have freedom of speech if, through dismissal, we remove the freedom to teach and thus speak?

To allow a union in concert with an administration to compel adherence to the dictates of a contract, which if not lived up to results in one's dismissal, in my opinion does abridge the freedom of expression.

After the first year the union was formed, there were 45 faculty persons who chose and were allowed what the union referred to as conscientious objector status. The following year the union unilaterally chose to remove the conscientious objector option, asserting that unless one became a member or paid service fees, they would be dismissed. All but two faculty members gave up their conscientious objector status. Intimidation can work sometimes...for a while. The union attempted to have the university administration dismiss the other two faculty persons, but the administration refused.

The hostilities between the leadership of the union and the administration had, in my opinion, not totally abated, and it seemed to be advantageous to the administration not to dismiss the faculty members, thereby forcing the union into what some might consider an embarrassing position, i.e., having to sue the administration to get them to fire a faculty member for whom the union allegedly was formed to protect.

Both faculty members were thus given the option of sending money to the WMU-AAUP chapter earmarked to go either to the National AAUP Chapter's Freedom Fund or the WMU Student Scholarship Fund. (How could anyone refuse to support freedom and scholarship?) In my opinion, however, it was not freedom and scholarship that was being fostered by this move, but rather it was more toward the very conformity which was originally sufficiently abhorrent to the faculty so as to form the union in the first place.

Since such a possibility seemed to be the case and since I had already sent the specified amount of money to an organization of my choice, I refused to conform. During all this time, I received what I construed as pressure to give in and conform, i.e., letters, phone calls, personal confrontations and arguments from union members. I finally received a letter in April, which was at least three months after the original notice from the WMU-AAUP union, which stated:

> "In view of the uncertain status of communications between you and the Chapter during the Fall, 1977, semester, the Executive Committee has voted to accept your proposal that a contribution to Rolling Thunder equal to the amount of the service fee be considered as satisfying contract requirements for this year only."

Rightly or wrongly, I considered the stated reason for this change in the union leaders' strategy to be less than honest. Indeed, as the intensity of their resolve to force me into conformity with union dictates against my will increased, it seemed more imperative than ever that I express myself clearly in dissent.

Mr. Scopes in the famous "Monkey Trial" was defended because the founder of the ACLU felt that such a public confrontation was the only way to dramatize the problem at hand. A school should not be able to fire someone who was expressing his belief, i.e., Darwin's theory of evolution.

At WMU in 1979, my colleagues and I can say verbally almost anything we like...if we pay for the opportunity to do so. "Pay to whom we dictate, Roger Ulrich, and you can say that you feel it's wrong for the AAUP to intimidate faculty members into assuming a position against which they might otherwise stand." Is it the case, I wonder, that nowadays freedom of speech can be bought by money on our university campuses as well as elsewhere?

Failure to pay either the AAUP National Freedom Fund or the WMU Student Scholarship Fund means that I will be "fired." My students in my ethics class, PSY 650, will be taught by someone else, as well as the students in my other classes. My appointment to do research and report my findings will no longer be supported because the AAUP and the administration have agreed that my freedom of expression has limits that are tied to the making of a payment of money wherever they decree.

As I have stated earlier, it was the fact that the faculty union seemed only interested in more dollars in their pockets that turned me from supporting them in the first place. Had they been concerned with convincing the administration that their hunger for money at the expense of research and teaching and, generally speaking, the students' welfare was a more important issue, I may have joined them and they wouldn't have had to try to force me into conforming by threatening me with dismissal. Indeed, I was a member of the local AAUP chapter but resigned because of what I am telling you now.

Let's look at another issue in which the ACLU did become involved. In a sense there was a union of Skokie citizens who wanted conformity just like the faculty union at WMU. They seemed to be saying "join our belief system, which says you cannot gather and say words on our Skokie campus such as 'Heil Hitler.'" I think it was true that the Nazis wouldn't have been able to gather and speak their thoughts had not we of the ACLU, in a sense, paid their dues for them. We intervened and supplied our muscle to help buy them their freedom of speech.

I am not suggesting that we shouldn't have done that. I'm also not going to change my stance and pay either the AAUP or Western Michigan University for my freedom of speech in the manner that is presently being promoted. The contract that was negotiated goes against every promise that was ever made to us as members of the academic community with tenure. "You should be dismissed only for cause, certainly not for failing to kick back some money to your bosses" *(who, in the case of the WMU faculty union, is not even contending that the kickback go to them as a share of the costs of negotiation)—see ACLU Policy #46.*

It is my opinion that the resolve of both the faculty and administrative leadership at WMU to stand for academic freedom has buckled under the dollar sign and the desire for conformity.

Maybe if I were really interested in getting the ACLU to support me, I should call myself a Nazi. My name is right for it. Then I could complain that I'm not being allowed freedom to say "Heil **(I'm not in sympathy with the demands of the AAUP)** *Hitler," not in Skokie but in Kalamazoo at WMU. But, of course, that would be silly, and I only said it to try to make a point. Our support of the Nazis is not understandable in light of our lack of support for this issue. I am, of course, not sure that what I am doing is right in any ultimate sense, but as I said before it seems to me that freedom of speech is freedom of expression, and our words are not the only thing that can be used to express oneself. Money also talks—certainly every citizen of the U.S. has heard that expression.*

Some folks are paying and will remain free to express themselves in WMU classrooms. By not paying, I am being told that I will not be allowed to express myself any longer in WMU classrooms.

ACLU Policy #46 says that the principle of freedom of association includes the freedom not to associate. Should we add that the principle of freedom of speech includes the freedom not to speak? Or, more to the point of my situation, the freedom to pay to the AAUP membership or service fees, or the AAUP National Freedom Fund, or the WMU Student Scholarship Fund, includes the freedom **not** *to pay to the AAUP membership or service fees, or the AAUP National Freedom Fund, or the WMU Student Scholarship Fund?*

I guess for those persons in our ACLU who see that there is a substantial and proper analogy between labor unions and government but they stop there, then the decision that we cannot support the part of us that is me can be understood. It is true, I believe, that it is generally accepted that a citizen who receives the benefits of a democratic government should share in its costs. The question that I always thought we members of the ACLU asked had to do with the **perception of benefit** *and how those "benefits" affect one's civil liberties. I may see no sense nor wish to help pay for the continuance of a group that negotiates away my* freedom to dissent, but I will continue to do my best to remain compassionate to their efforts to try. Certainly the leadership of the ACLU has every right not to fight for my particular cause, although it strikes me as being at least as important as whether or not an older lady teacher can force a younger lady student into wearing a bra! Some years back, before it became a big deal in Portage, my then-12-year-old daughter was kicked out of junior high school for not having hers on. She didn't have any reason to have it on either, since she owned neither a bra nor breasts.

I went up to school with a lady friend of mine and tried to comprehend the thinking of the school principal and his assistants. My lady friend asked him if he were offended by the fact that she didn't have a bra on. She also asked him whether or not he had considered making all the boys wear jocks.

Well, to make a long story short, I left the meeting sort of unhappy, but my daughter calmed me down and told me to forget it: "There are more important battles to be fought."

I think she was right, and the issue at stake at WMU as it relates to the willingness to fire a tenured faculty member for refusal to kneel to the dictates of the union-administration coalition, which says do as we say or get fired, is one.

The local ACLU has chosen not to support me. Given that many of them are members of the faculty union makes sense of the fact that they were not interested in my case.

I have always assumed that it is pretty important that we watch very carefully what we join and how tightly we join it if we are to remain flexible in regard to the civil liberties of ourselves as well as others.

I guess I'm learning to be more wary of unions, be they labor, AAUP, American Civil Liberties, or the union between judges and attorneys who seem to forget at times that human laws are meant to be broken. Only natural laws remain, and they are meant to be understood. If my profession as an explorer of human nature and the natural laws which surround it will not be tolerated at WMU, and my expression, either verbal or nonverbal, of the results of my findings is forbidden, then I will go elsewhere.

In closing, let me wish you well; keep up the good work whenever you can and, for goodness sakes, do keep Policy #46 under review, because it sure needs it!!! And if you don't think it needs review, then take it to the traditional Indians of the American Indian Movement and ask *them if they feel that because they have now been made citizens of the U.S. they should share in its costs, especially those costs that are incurred as a function of the continued perpetuation of genocide within the Native communities.*

With respect,
Roger E. Ulrich

The Honor of Samuel I. Clark

April 4, 1979
Dr. John T. Bernhard, President
Western Michigan University
and
Dr. Don R. Lick, President
WMU Chapter, American Association of University Professors

Dear Presidents Bernhard and Lick:

Dismissing Professor Roger Ulrich from the University because he has failed to pay his AAUP dues (or otherwise comply with the contract between the AAUP and WMU regarding to whom an equivalent sum of money should be paid) will likely evolve into a major error of the University.

Roger was first employed at the University in 1965. He is a tenured Research Professor. No question is raised about his competency. The sole issue in the matter is his failure to conform to a clause in the contract, negotiated between the University and the WMU Chapter of AAUP, stipulating that a substitute payment for dues be paid as specified by the contract or chapter. This Roger Ulrich does not do, objecting as a matter of conscience.

The law is fickle, and the U.S. Constitutional prohibition that "no state shall pass any law impairing the obligation of contract" may well be judged inapplicable, even though Western Michigan University is a unit of the State of Michigan. However, the University is party to a contract with Roger Ulrich, "negotiated" prior to the current contract between the AAUP and WMU. Is it proper or possible that a contract between the University and one faculty member initiated in 1965 can be terminated by a subsequent contract between the University and other faculty in 1978? Indeed is the University free to contract with third parties for the termination of obligations agreed to earlier?

Roger Ulrich has or had a right which many faculty believe they also have—the right to tenured employment. This propertied right may be taken from him through a commitment the University has made to the WMU Chapter of AAUP. The University should not do this, and AAUP should not ask the University to do it.

However, it is the University who will terminate the employment of Roger Ulrich, and it is the University who will be the defendant in litigation. And it will be the University who is faulted for diminishing academic freedom.

Recent litigation ran against the University when the number of years-in-service were changed for purposes of determining tenure. These relatively moderate alternatives in the conditions of employment, which were made after surmised contractual understandings occurred, were vigorously criticized by many members of AAUP. And the settlements out of court to faculty thus injured were supported by members of AAUP. It is rationally unacceptable to now have a complete reversal of argument. What presumably was wrong for the University to do cannot now be right simply because it enforces the chapter's power.

Roger argues that he is denied freedom of speech through the collusion of the University and AAUP. At first the argument may seem remote. Two things make it real. First, there is Roger's interest as a psychologist in nonverbal communication—his actions like gestures are speech. (Who would maintain that the deaf and dumb have truncated First Amendment rights?) Second, there is an obvious effort to "subdue" Roger undertaken by AAUP. While I do appreciate the embarrassment AAUP must be experiencing, nevertheless they are fixed on conquering Roger's stubbornness. Last year they agreed to Roger's payment to Rolling Thunder. Presumably Roger would make a similar payment this year.

It seems clear that Roger and the AAUP are both living out roles and thus proceeding on principle: Roger walking away from a situation which he judges has no right to intrude upon his autonomy, and the AAUP asserting its authority to achieve some sign of submission. Were Roger more formally courteous or AAUP not so tendentiously persistent, the disagreement would pass away. If it does not pass away, Roger's departure will hurt the University and the AAUP more severely than himself.

There are several solutions. The best is to drop the entire matter. Let Roger pay whom he will, AAUP withdrawing its request of March 7, 1979, to the University.

Next is the route of a negotiated solution. Have someone be a go-between. Nothing irreconcilable exists in the disagreement.

Third is the refusal of the University to execute the AAUP request of March 7, 1979, on the essentially technical question that past practice with Roger was that of permitting payments to persons of Roger's selection.

Most surely one thing should not happen in this conflict of wills and principles—the dismissal of Roger Ulrich. It is unconscionable to do a serious harm saying that nothing personal is intended or that only law and contract are being secured. Crimes are committed every day in the guise of law. Histrionic acts are often horrors.

Sincerely,
Samuel I. Clark
Director of Honors College

Students Pay for Professor's Union Dues
by Kevin Bramble, Associate News Editor

The following article appeared in WMU's student newspaper, the Western Herald, *on May 14, 1979.*

Efforts to terminate Western Michigan University psychology professor Roger Ulrich's contract by the American Association of University Professors (AAUP) because of his refusal to pay dues have been halted, at least temporarily, by a coalition of WMU students who have raised the necessary funds to pay Ulrich's union expenses.

The student group, called the Students for Academic Freedom, organized after grievance officer Tom Mannix had declared Ulrich in violation of the union's contract with the university. At the April 5 hearing, which was requested by the local chapter of the AAUP, Mannix stated that the university "agreed as management to terminate employment of anybody not complying with the shop provision."

The money contributed by the Students for Academic Freedom marks the three-year point in the Ulrich-AAUP struggle. After the faculty voted to accept the local AAUP chapter as their official bargaining agent in 1976, Ulrich filed for conscientious objector status with the union.

"For a variety of reasons, I had not been especially impressed with the chapter's organizing efforts," explained Ulrich. "For a long time I have been involved in a personal campaign to try to live on less money and to waste fewer of our Earth's resources. Those organizing the union seemed mainly concerned with forcing an overpaid administration to make across-the-board salary hikes to all faculty. It seems like we all get caught in the over-consumer trap."

Ulrich also opposed the AAUP's effort to include a clause in their contract with the WMU faculty which "would allow them to threaten dismissal to anyone who did not conform to their demands for a monetary kickback in the form of mandatory union dues."

Ulrich wrote in an open letter, "It appeared that a group whom I had considered for years to be libertarian, and who would fight any and all who would use the threat of dismissal to force compliance to any cause which threatened the academic freedom of its members, was now willing to use the same strategy to force compliance from their own members.

"They were supposed to protect you so you could be a Nazi, a Communist or a dissenter of some sort and still teach."

As defined in the conscientious objector clause in 1976, Ulrich agreed to donate an amount of money equal to his union dues to the Rolling Thunder Indian Charity. "Sending that amount of money to a cause of my choice I saw as a compromise which allowed me to show it was not the loss of money per se to which I objected," stated Ulrich.

The union accepted Ulrich's conscientious objector stance and his donation to the Indian fund.

However, the following year the union reversed its position regarding conscientious objection status. The donation to Rolling Thunder was no longer accepted as an alternative to paying union dues. Ulrich was told he must pay membership dues or service fees. "I told them over and over again that I wasn't going to pay," said Ulrich.

"The union was willing to eliminate all options other than the one preferred by them and to force compliance to its will by threatening dismissal to those who would not conform."

Ulrich again sent a donation to Rolling Thunder instead of paying union dues. The AAUP executive committee agreed to accept his contribution instead of union dues and thus Ulrich had satisfied contract requirements for the 1977-78 school year.

However, this was the last time the union allowed a conscientious objector to donate money to a cause of his or her choice instead of paying union dues. The AAUP's new position regarding conscientious objectors limits dissenters to donating to two groups: the AAUP Academic Freedom Fund or the university's student scholarship fund. Thus, Ulrich's request to contribute to the same Indian organization again this year was denied by the union.

According to AAUP President Don Lick, the denial of Ulrich's request is in congruence with the new agreement between the faculty and administration. Lick said the AAUP reestablished their position because "it was felt that too many people were trying to take advantage of the conscientious objector clause."

Ulrich will not abide by the new clause set forth by the AAUP. "I am not willing to send money to either the AAUP National Freedom Fund nor the Western Michigan University Scholarship Fund," stated Ulrich. "I am unwilling to reward either of the parties who have conspired together on this issue, and who together have agreed to force the dismissal of anyone who fails to conform."

"There may be some people in the membership who don't like him," said Lick, "but the union leadership is not out to get Roger Ulrich. When you have a contract you have to abide by it."

With the intervention of the Students for Academic Freedom, Ulrich's union dues have been paid. However, the conflict between Ulrich and the AAUP is far from being solved. "All this does is delay the problem for a year," said Lick.

Suspension

March 6, 1981
Dr. John T. Bernhard, President
Western Michigan University

Dear Dr. Bernhard:
 Enclosed are copies of correspondence I received from Dinah J. Rank, Director of Collective Bargaining and Contract Administration.
 In short, I am to be suspended without pay on Thursday and Friday, April 23 and 24, 1981. The reason for my suspension, although complex,

is basically for failing to conform to the collective bargaining agreement between Western Michigan and the WMU/AAUP chapter. My failure to conform is based on: (1) My feeling that the WMU/AAUP chapter in organizing was basically motivated by greed for money similar to that which motivated the WMU administration to continue increasing their personal salaries. See the first page of my letter to the AAUP dated October 5, 1976. (2) My feeling that one deserves at a University the right to freedom of speech without undue threat of reprisal. (3) My feeling that the administration at Western Michigan University has failed in their obligation to guard against such threats and indeed has instead chosen to collaborate as witnessed to by my suspension. (4) The conviction that the fiscal policies you both promote are contrary to known facts relative to our limits for growth and would lead to severe economic problems. In the University Assembly Report reported in the WMU Western News extra of February 3, 1981, there is contained remarks by you, Vice President Wetnight, Vice President Ehrle and Maury E. Parfet, which attest to your current conviction that the economic situation in the nation, state and at WMU do indeed constitute a problem. You, in fact, refer to them yourself as "our fiscal crisis."

I respectfully request that you personally remove my suspension. It makes little sense to suspend me for predicting the problems that are now upon us not only with words but a willingness to not be intimidated into acting against my convictions (that what I said was true) by threats from an administration who is now asking that we trim the same sails that I called for trimming in 1976. When you suspend me for my unwillingness to join forces with those who were leading us toward the necessity to "borrow from the future," perhaps the governor should consider suspending Maury E. Parfet as chairperson of the Board of Trustees, who now says we must not "deceive ourselves about the gravity of our situation, or plan on false hope." And when Governor Milliken considers that possibility, he should then consider suspending himself for two days without pay as leader of the system which has arranged to suspend me.

I respectfully request that you consider what I suggest and why I should be suspended and not you and the others I have recommended.

Roger Ulrich
Professor, Department of Psychology

Prof Faces Two-Day Suspension for Failure to Pay Union Dues
by Ed Stych, Staff Writer

The following article appeared in the April 8, 1981, edition of the Western Herald.

Roger Ulrich, a research professor in psychology at Western Michigan University, has been given a two-day suspension without pay for not paying dues to the WMU Chapter of the American Association of University Professors (AAUP).

The suspension will take place on April 23 and 24, 1981, and, according to Ulrich, he will lose about $300. He was suspended two days last year for the same reason.

Ulrich is in non-compliance with Article IX, Agency Shop, of the collective bargaining agreement between Western and the AAUP, which was modified in 1979.

Article IX states that "all persons in the bargaining unit shall tender payment to the Chapter (AAUP) of either the dues assessed on members, or the service fees assessed on non-members, or, if granted conscientious objector status, of an amount of money equal to the annual service fees assessed on the non-members to be forwarded intact to the WMU Scholarship Fund or the AAUP Academic Freedom Fund."

Faculty who refuse to abide by Article IX are to be suspended for two days without pay. Ulrich has been granted conscientious objector status, but has refused to pay to any of the choices Article IX gives him.

His reasons for not joining the union were first voiced in 1976, when the AAUP, which was originally formed to protect professors' academic freedom, voted by a narrow margin to become a bargaining agent for the faculty.

Ulrich, who has been teaching at Western since 1965, had been a member of the AAUP before 1976 because he believed that they were the group most concerned with guarding his freedom to teach.

"But now it seemed the AAUP was standing not so much for my freedom as a behavioral scientist, but something for which I had less interest...a struggle for more money for a group who was already well off," he wrote in an open letter in 1976.

In the 1976 letter, Ulrich called for salary cuts instead of raises, and wondered if the faculty was spending too much time getting rich and not enough time doing research and excellent teaching.

Ulrich refuses to pay because he does not believe he is reaping any benefits from the AAUP, since he is against salary hikes and other benefits the union fights for.

But the AAUP has a different opinion.

In a 1976 letter, the AAUP said, "We find no philosophical difficulties with the proposition that all should pay a fair share...those who reap the benefits, along with active members, should pay their fair share."

"His (Ulrich's) opinions are not the same as ours," said George Miller, president of the AAUP at Western. "The Agency Shop article in our contract says he has to pay his dues or a service fee, or be suspended."

Miller said that Ulrich has never turned down a pay raise or any benefits. Also, the money taken out of Ulrich's salary for the suspension, which is considerably higher than union dues, will be sent to either the WMU Scholarship Fund or the AAUP Academic Freedom Fund.

According to the American Civil Liberties Union (ACLU), the AAUP has every right to force Ulrich to pay. The following was taken from the ACLU's Union Shop Policy.

"It is generally accepted that a citizen who receives the benefits of democratic government should share in its costs; similarly, a citizen of industrial government (which a labor union is considered to be an instrument of) cannot claim a right not to contribute to its costs."

The ACLU goes on to say that a worker does not have to do any more than contribute to the costs of representation. "The principle of freedom of association includes the freedom not to associate." The policy is currently under review by the ACLU.

The administration at WMU feels they should support the contract they signed with the AAUP in 1979.

Martin Gagie, executive assistant to the president, said President Bernhard refused to remove Ulrich's suspension because their legal advisers said they had to support the contract.

When asked if the administration supported the Agency Shop article, Gagie only said, "We support the contract." He did say, however, that Article IX could be a topic for discussion during the next contract negotiations with the AAUP.

Ulrich and the AAUP have been struggling with each other since 1976. For the 1976-77 school year, he was given conscientious objector status and was allowed to send an amount of money equal to the service fee to a cause of his choice, which was an American Indian group.

"This allowed me to show it was not the loss of money per se to which I objected," said Ulrich.

The contract in effect between 1976 and 1979 stated that refusal to pay any money at all would result in dismissal instead of suspension.

For the 1977-78 school year, the AAUP dropped the conscientious objector clause, forcing everyone to pay directly to the AAUP. Ulrich refused to, and was told he would be dismissed. However, after Ulrich sent another donation to the Indian group, the AAUP accepted it as satisfying contract requirements.

The conscientious objector clause was reinstated for the 1978-79 school year. This time, however, the money had to be sent to either the WMU Scholarship Fund or the AAUP Academic Freedom Fund. Ulrich refused and was told he would be dismissed at the end of the winter semester.

However, just before he was to be fired, a group called Students for Academic Freedom raised over $150 to pay for his service fees. The union's request for his dismissal was then dropped.

"I'm pretty sure I wouldn't have been around by then if I wasn't tenured," said Ulrich. He also said that there are many other teachers who are against the AAUP, but won't fight it.

The suspension clause was put in place of the dismissal clause in the Agency Shop section of the contract for the 1979-80 school year.

"We feel this is more legitimate and more just," said Miller.

Last year, just as this year, Ulrich was suspended for two days. Although he was not required to, he worked both days last year and says he will this year, too.

Ulrich says he expects to be suspended every year now, since he still refuses to join the AAUP.

Animal Research: A Reflective Analysis

Reprinted from Psychological Science, *Vol. 3, No. 6, November 1992, pp. 384-386*

The following is a reflective commentary on Miller's (1991) "Commentary on Ulrich: Need to Check Truthfulness of Statements by Opponents of Animal Research," which addressed my 1991 commentary "Animal Rights, Animal Wrongs and the Question of Balance."

In a 1948 paper, Neal Miller reported his conclusion that he had trained rats to fight by removing shock each time the animals approximated the now-familiar reflexive fighting position. He concluded that this positioning was an escape reaction reinforced by the termination of electric shock. Following my attempts and those of N.H. Azrin, Don Hake, Bill Holz, Ron Hutchinson and others at the Behavior Research Laboratory at Anna State Hospital, Anna, Illinois, to replicate Miller's procedures, we concluded that this positioning, in fact, was a natural reaction to electric shock and had little to do with operant conditioning (Ulrich & Azrin, 1962). Miller's interpretation of his results was different from ours, perhaps due to his theoretical focus and his interest in linking psychoanalytic displacement and stimulus–response generalization. So it was that an experiment involving pain-producing shock in Miller's lab led to our shocking still more animals.

In subsequent reviews of the literature, we further discovered that O'Kelly and Steckle (1939) had reported findings similar to ours in an article titled "A Note on Long-Enduring Emotional Response in the Rat." Several other early papers had also made the same observations with rats, including one by Daniel (1943) and one by Richter (1950). So, in "Reflexive Fighting in Response to Aversive Stimulation," published in the 1962 *Journal of the Experimental Analysis of Behavior,* Azrin and I "proved" once again that animals will fight when wounded. When I told my Mennonite mother what I had discovered in my dissertation, she said, "Well, we know that. Dad always told us to stay away from wounded animals on the farm because they might hurt you."

Nevertheless, for the next 10 years I would demonstrate in countless ways that animals are more likely to be aggressive when they are hurt. I wrote on aggression and even assisted the production of movies dealing with aggression. I helped design new strategies and equipment for shocking a multitude of creatures. I even convinced children to shock some rats and "watch what happens."

It was about 1972 when I seriously confronted the "painful" truth: my research had merely demonstrated an already well-known phenomenon. Also, I concluded that the "findings" had not contributed significantly to an understanding of human aggression since they were artifacts produced artificially from an unnatural setting, more often than not completely irrelevant to the prediction and control of most human aggression. Indeed, the conditions under which the animals were painfully stimulated never really mirrored the human conditions about which I presumed to generalize.

How had I become so deeply involved in the enterprise? I suggest in the same way as had Miller and all other animal researchers: animal experimentation is self-reinforcing! Each study leads to more studies that lead to more trips to conventions, more presentations, more publications, tenure advancement, more attention. Every once in a while we find something truly amazing which, no doubt, helps relieve a human problem for a while. However, more and more, the question arises: *should we continue doing research with animals that often shows little regard to possible harmful effects for the future survival of all life*? Unfortunately, the voluminous literature we generate is continuing to grow, with no end in sight.

Let me cite another example that I see as contributing to the problem of animal exploitation. Recently a colleague who, I feel, tends to believe that all behavior can be explained according to operant principles wrote in the newsletter of the Association for Applied Behavior Analysis:

> What about aggressive reinforcers rather than aggressive behavior? Perhaps the big deal isn't the electrically-shocked monkey biting the rubber tube. Perhaps the big deal is the possibility that pressure on the monkey's teeth will reinforce whatever arbitrary response produces that pressure (Malott, 1990, p. 1).

Often my behaviorist colleagues criticize theorizing about non-observables as cognitive thinking. Yet it is what I consider to be such off-hand theorizing that eventually causes animals to again be subjected to cruel and painful conditions. To find out if my colleague is "correct," someone (probably a graduate student needing a thesis topic) will conduct yet more research. At no small cost in animal suffering, money and time, those who might pursue his question, of whether the painful shock or the sensation of biting reinforces aggressive behavior, will perpetuate the overall problem of needlessly-cruel animal exploitation.

My early career in animal research, like that of many scientists, including, I feel, Dr. Miller, shows a tragic irony. While artificially inducing various behaviors, including aggression, in coerced animals, we—not our subjects—were and are the true exemplar of human aggression. For many years I tormented, injured and killed laboratory rats bred to be docile. I was, in fact, responsible for so much animal suffering that my research was identified by one animal protectionist as an example of cruelty (Rowan, 1984).

Was my research worthless? Was Miller's research worthless? It depends on your point of view. Is there, as Miller (1991) has suggested, a "long history of false and extremely misleading statements by opponents of animal research" (p. 422)? And are animal researchers equally vulnerable to the same criticism? It depends, again in my view, on who is doing the "reflective analysis."

While I now question that my data on "aggression" in rats shed much new light on human aggression, I believe my personal experiences in the laboratory represent a useful case study of human aggression. B.F. Skinner, although a behaviorist, recognized the importance of self-reflection in the study of human psychology. In *Walden Two,* Skinner's (1948) fictional student, discussing the importance of psychology, tells his professor, "It's a job for research, but not the kind you can do in a university, or in a laboratory anywhere. I mean you've got to experiment, and *experiment with your own life!* Not just sit back—not just sit back in an ivory tower somewhere—as if your own life weren't all mixed up in it" (p. 5).

However, many of Skinner's followers have focused their research and writing on behavioristic assumptions, which I feel ignore his important insight on the nature of self-reflection. In contrast, Don Bannister (1981) has repeatedly called on his colleagues to use self-reflection in the study of psychology:

> There is a growing recognition in psychology generally that we may have to abandon our simple mimicry of the natural scientist and recognize that we cannot usefully experiment *on* our subjects, we have to experiment *with* them. Experimenting *on* animals offers us a way of delaying the day of that recognition.
>
> We can now begin to see that in psychology the ethical issues involved in experimenting on animals are not separate from the scientific issues. It is not simply being argued here that it is an unkind practice to experiment on animals and that at the same time it happens to be a not very clever practice. The unkindness and foolishness stem from the same source, that is, from a particular notion of what 'being a scientist' is about in psychology. If psychologists continue to believe that a 'scientific' psychology must be 'objective' in the manipulative and non-reflexive sense of that term, then they will use those strategies which favor that kind of 'science'. Animal experimentation is such a strategy. It allows the psychologist to ride on the back of existing cultural and ethical permissions about, and gulfs between, species.
>
> Picture the kind of psychologist who sits, notebook in hand, watching rats drown in a water-filled glass maze as they desperately strive to find the exit, thereby increasing his or her knowledge of 'learning under stress'. He or she is not simply personally indifferent to suffering but is trapped, as surely as the rat, within a total view of science and of his or her own nature as a scientist. Psychology has failed in that it has given such a person no psychological view of his or her own character by the nature of his or her predicament (p. 316).

I still believe that psychology is an important field of study, and I believe that *Psychological Science* is doing a profoundly important job in airing dialogues, such as mine and Miller's, concerning an extremely complicated question, that is, the value of animal research in relation to the long-term survival of all life on Earth. However, as Skinner observed, eventually you have to "experiment with your own life." Each human thought and action is a manifestation of human psychology, from falling in love to shocking rats. By using ourselves and our own lives as the subject matter of our experiments, we may increase our understanding of human psychology. We know that humans are capable of extremely aggressive and selfish behavior, and our self-indulgence has created obscene inequalities of wealth. We have wreaked ecological havoc, which has caused the extinction of countless other species and now threatens our own. For years I studied aggression, and I have come to the conclusion that the first step in reducing human aggression is to take personal responsibility and reduce it in ourselves. It is in this spirit that I respectfully respond to Miller's "Commentary on Ulrich."

Finally, I am not an opponent of animal research, and I continue to plead for an open discussion of its relevance to the human situation with wisdom and balance, and with respect for the never-ending search for truth. This is a position upon which Miller and I agree, and a position which I pray will serve as a touchstone for further dialogue within the total human community.

References

Bannister, D. (1981). The fallacy of animal experimentation in psychology. In D. Sperlinger (ed.), *Animals in research: New perspectives in animal experimentation.* New York: Wiley.

Daniel, W.J. (1943). An experimental note on the O'Kelly-Steckle reaction. *Journal of Comparative Psychology, 35,* pp. 267-268.

Malott, R.W. (1990). Notes from a radical behaviorist. *The ABA Newsletter 13,* p. 1.

Miller, N.E. (1948). Theory and experiment relating psychoanalytic displacement to stimulus-response generalization. *Journal of Abnormal and Social Psychology, 43,* pp. 155-178.

Miller, N.E. (1991). Commentary on Ulrich: Need to check truthfulness of statements by opponents of animal research. *Psychological Science, 2,* pp. 422-424.

O'Kelly, L.E., & Steckle, L.C. (1939). A note on long-enduring emotional response in the rat. *Journal of Psychology, 8,* pp. 125-131.

Richter, C.P. (1950). Domestication of the Norway rat and its implications for the problem of stress. *Associated Research in Nervous and Mental Disease Processes, 29,* p. 19.

Rowan, A.N. (1984). *Of mice, models & men.* Albany, NY: SUNY Press.

Skinner, B.F. (1948). *Walden Two*. New York: Macmillan.

Ulrich, R.E. (1991). Animal rights, animal wrongs and the question of balance. *Psychological Science, 2,* pp. 197-201.

Ulrich, R.E. & Azrin, N.H. (1962). Reflexive fighting in response to aversive stimulation. *Journal of the Experimental Analysis of Behavior, 5,* pp. 511-520.

Animal Research: A Psychological Ritual

Reprinted from The Animals' Agenda, *May 1991, pp. 40-44*

A fundamental assumption of the science of behavior is that studies of nonhuman animals can yield results that ultimately benefit humans. From Ivan Pavlov's dog-conditioning experiments to the present day, animal researchers have generally assumed the view articulated by the late B.F. Skinner in 1953:

> We study the behavior of [nonhuman] animals because it is simpler. Basic processes are revealed more easily and can be recorded over longer periods of time. Our observations are not complicated by the social relations between subject and experimenter. Conditions may be better controlled. We may arrange genetic histories to control certain variables and special life histories to control others—for example, if we are interested in how an organism learns to see, we can raise an animal in darkness until the experiment is begun. We are also able to control current circumstances to an extent not easily realized in human behavior—for example, we can vary states of deprivation over wide ranges. These are advantages which should not be dismissed on the *a priori* contention that human behavior is inevitably set apart as a separate field.

The use of animals in behavioral studies is built upon such assumptions and has evolved into a technology practiced mainly for the purpose of proving them.

A conflicting assumption, again skillfully articulated by Skinner in the novel *Walden Two,* is that ultimately there is no experiment other than a real situation:

> "Some of us feel that we can eventually find the answer in teaching and research," said Professor Burris.
>
> "In teaching, no. It's all right to stir people up, get them interested. That's better than nothing. But in the long run you're only passing the buck—if you see what I mean, sir." Rogers, the former student, paused in embarrassment.
>
> "For heaven's sake, don't apologize," replied Professor Burris. "You can't hurt me there, that's not my Achilles heel."
>
> "What I mean, sir, is you've got to do the job yourself if it's ever going to be done, not just whip somebody else up to it. Maybe in your research you are getting close to the answer. I wouldn't know."
>
> "I'm afraid the answer is still a long way off," Burris demurred.
>
> "Well, that's what I mean, sir. It's a job for research, but not the kind you can do in a university, or a laboratory anywhere. I mean you've got to experiment and experiment with your own life, not just sit back in an ivory tower somewhere—as if your own life weren't all mixed up in it." Rogers stopped again.
>
> "Perhaps this was my Achilles heel," said Burris.

The contrived basic experimental laboratory that has evolved from Pavlov's work and the real-life application of knowledge are in fundamental conflict, a conflict increasingly evident from the failure of behavioral science to effectively respond to challenges including urban alienation, violent crime, child abuse, substance abuse, the continuing proliferation of age-old forms of mental illness and what often seems to be a complete collapse of the elementary and secondary levels of our educational infrastructure. Effective responses in many cases have long since evolved, mainly at the clinical, police beat or grade school classroom level; but at the academic level, where most of the federal mental health budget is spent, the emphasis is on research—mainly animal research—and the best minds in the behavioral field are continually directed into research, away from actual prevention and cure.

The contrived basic experimental laboratory

In June of 1961, I completed my doctoral dissertation entitled "Reflexive Fighting in Response to Aversive Stimulation." The study, which involved shocking rats, showed that stereotyped fighting would occur between paired animals as a reflex-type reaction to pain prior to any specific conditioning. My paper was later published in the *Journal of the Experimental Analysis of Behavior.*

I was close to obtaining a Ph.D. in clinical/counseling psychology from Southern Illinois University, which was at that time trying to win American Psychological Association recognition for producing clinicians who were also scientists. Behind that effort was the inferiority complex felt by many clinical psychologists in the face of the American Medical Association and its psychiatrists. Only research with "quantifiable" data was acceptable to dissertation committees, whose basic behaviorist assumptions didn't allow for the contemplation of such variables as emotions, feelings, disgust, etc., nor for questioning why one was shocking rats in the first place.

Simultaneous with various animal research projects, I was also conducting studies with mental patients. My research with patients, however, was often looked upon by those who held radical behaviorist views as being too complex to allow for "clean data."

For me, the scientific attraction to animal research had, in the final analysis, little to do with a demonstrable relationship of research findings to the goal of "helping humans." In retrospect, I would say the main attraction to working with animals was, as Skinner proclaimed, "that we are able to control current circumstances to an extent not easily realized in human behavior." At any rate, after I earned my Ph.D., I joined the army of animal researchers who contended that we must conduct further experiments.

Laboratory aggression experiments provide a perfect example of basic research, in which the sequence of events leads from one animal experiment to the next, with each project following the preceding one as a direct consequence, and with each being essentially as irrelevant to solving real human problems as the one before. The fact that I have often sat behind closed doors with numerous colleagues who have agreed with this analysis is of little consequence to the animals still confined in laboratory cages around the world, because the true feelings of these professionals remain unexpressed.

Let us look beyond closed doors, however, at some additional data from the "contrived" research situation.

In 1948, a study was published by Neal E. Miller under the title "Theory and experiment relating psychoanalytic displacement to stimulus-response generalization." It is a report of how Miller and his assistants trained rats to fight by removing the shock each time the animals approximated the fighting position. Fighting, they presumed, was an escape reaction, reinforced by the termination of electric shock. Our laboratory at Anna State Hospital was very much involved at the time in escape and avoidance research and was especially concerned with the area of punishment. An attempt to replicate Miller's procedures, however, showed that fighting behavior could be elicited from paired rats with no training whatsoever. Here, now, was a perfect example showing not only that the Miller interpretation of results was incorrect, but also that a wrong-headed analysis was being used by people dedicated to using only "observable data."

This, by the way, is the same Miller who later offered an extensive defense of animal research in psychology in a 1985 article whose abstract cites the following list of animal research benefits:

> ...Treatment of human urinary and fecal incontinence; psychotherapy and especially behavior therapy and behavior medicine; rehabilitation of neuromuscular disorders; understanding and alleviating effects of stress and pain; discovery and testing of drugs for treatment of anxiety, psychosis, and Parkinson's disease; new knowledge about mechanisms of drug addiction, relapse, and damage to the fetus; treatment enabling extremely premature infants to gain 47% more weight and save $6000 per child in hospital care; and understanding mechanisms and probable future alleviation of some deficits of memory that occur with aging.

I am essentially unimpressed with Miller's "faith list" defense of animal research benefits. This is especially true of his last point, concerning memory loss with aging. In a conversation with Dr. Miller in 1963 regarding his earlier experiment in which he allegedly taught rats to fight by removing shock, I remember that he recalled almost nothing about it.

So it was that a trivial experiment (albeit not so to the rats), done by a well-known apologist for animal research who had wrongly interpreted the results of his experiment, led to our shocking still more animals.

We, of course, went to the literature and somewhat unhappily discovered that what we had found had been found before by O'Kelly and Steckle in 1939. They titled their paper "A long enduring emotional response in the rat...," which no doubt it was and which it continues to be to this day, inasmuch as rats are still being shocked to demonstrate the pain-aggression phenomenon.

When I told my Mennonite mother what we had found in my dissertation research, she said, "Well, we know that. Dad always told us to stay away from wounded animals on the farm because they might hurt you." Nevertheless, I entered into a ten-year period of dedicated analysis of the causes of aggression, hoping a better understanding of how to control human aggression would follow.

In 1973 I finally came to the conclusion that if the control of human aggression was our goal, we were looking in the wrong place. I still was in no way enlightened in that area, to the extent that I could offer meaningful advice to people who questioned me regarding aggression. Indeed, my own anger was often uncontrollable, despite my discoveries and laboratory knowledge. Thus one spring, in response to my department chairman's question, "What is the most innovative thing that you have done professionally during the past year?" I replied, "Dear Dave, I've finally stopped torturing animals."

As early as 1972, I had already stopped conducting traditional basic animal research, having demonstrated over and over in countless different ways what my grandpa had taught his children: when animals are hurt, they are more likely to be aggressive. Without fully realizing it at the time, I was divorcing myself from the vast armada of behavioral scientists who daily illustrate how animal research has become for them a self-reinforcing activity.

For ten years I had written on the topic of aggression; did research; traveled through Europe, Asia, Central and South America, and the U.S. talking on the topic; made movies about it; wrote grant requests to every local, state and federal agency, private and public, that held even the remotest hope of giving money for my research. I helped design new strategies and new equipment for shocking anything that moved, and even observed children whom I had convinced to shock some rats and "watch what happens." More and more allegedly new discoveries were added to a voluminous literature, reprints of which I was collecting for a book and which now weigh close to 50 pounds. Studies leading to new studies, all involving countless animals, with the findings essentially irrelevant to people *in that at no time did the conditions under which the animals were studied equal the existing human conditions to which the generalizations were being "theoretically" transposed.* These permutations upon permutations, conducted in the world's scientific laboratories with different species under countless different research conditions, are nearly infinite.

The real-life laboratory

Skinnerites, perhaps more than any other group of scientists, have called for the generalization of animal research findings into building a better tomorrow. Their persistent claim is that experimental analysis of animal behavior has enabled us to redesign human culture to enhance our chances of survival. But for me, as for Skinner's hero in *Walden Two,* faith in the ability of animal research to guarantee the continuance of humankind on Earth is nothing less than pure superstition. Indeed, we are faced with a situation in which over 100 years of animal research may have left our culture further behind in the search for wisdom than when the research started.

In his book *Nature, Man and Woman,* Alan Watts summarizes:

> Based on the assumption that we had done wisely, and were still here and likely to remain, the human race had survived and seemed likely to go on surviving for perhaps more than a million years before the arrival of modern technology. We must on this premise assume that it had acted wisely thus far. We may argue that its life was not highly pleasant, but it is difficult to know what that means. The race was certainly pleased to go on living, for it did so.
>
> On the other hand, after a bare two centuries of industrial technology, the prospects of human survival are being quite seriously questioned. It is not unlikely that we may propagate, eat and possibly blow ourselves off the planet.

As is the tradition in science, I will now call for further research. But this is the question we must explore: can human society afford the assumption that the current level of animal research and sacrifice is worthy of our continued support?

My conclusion is no. The atrocities we persist in perpetrating within our laboratories, where scientists are paid to perform painful rituals on other life forms based on blind faith that human suffering might be driven away, should increasingly be questioned and discontinued. They are not reducing the suffering we so often feel and see around us in the real-life laboratory.

Our scientific addiction to animal research must be given up and replaced with the observation of natural phenomena. What B.F. Skinner said in his novel *Walden Two* about "the need for us to experiment with our own lives and not just sit back in an ivory tower somewhere—as if your own life weren't all mixed up in it," overshadows in importance every other point he ever made. If Skinner is to be remembered as an important voice in the history of science, it will be for his call to reconnect research with that which is relevant.

Animal Rights, Animal Wrongs and the Question of Balance

Reprinted from Psychological Science, Vol. 2, No. 3, May 1991, pp. 197-201

Abstract—The issue of "animal rights" is part of a larger, overriding problem of the rights of all living beings to a habitable Earth. The dangers posed to the overall environment by researchers who are insensitive to their animal subjects is not unlike that of animal rights activists who are equally unaware of the adverse effects which arise from some of their own daily actions. The present paper looks at both the world of science and antivivisection, making the case that all of us must begin to appreciate more fully the concept of there being limits to growth and the need to make better use of life-sustaining resources still available. Issues involved in the debate over ethics in animal research are extremely complex. Greater wisdom must be sought out and exercised in order to lead us toward policies that will work out best in the long run for both people and other animals. Scientists should not overreact to criticism of animal exploration and study which truly benefits the survival of all life on Earth. Furthermore, researchers cannot escape the logic that if the animals we observe are reasonable models of our own most intricate actions, then they must be respected as we would respect our own sensibilities. We cannot defend our scientific work with animals on the basis of the similarities between them and ourselves and then defend it morally on the basis of differences. Although many people criticize animal research in seemingly irrational ways, it behooves our society to note that increasing numbers of scientists as well as clinicians are criticizing animal research on scientific grounds. Finally, critics of the *animal rights movement, such as David Johnson, are not "wrong" for reporting their observations. Excesses are to be found, of course, just as they are found within the animal research institution and in our own daily lives. As scientists, we should look at all sides. The treatment of animals and humans, as well as all life, is a question of balance.*

In a similar vein to the problem of "animal rights" posed by insensitive researchers, a close analysis of the ethics espoused by some animal rights activists reveals actions as out of balance environmentally as are some of the behaviors observed in scientific laboratories. For example, a poster at the 1990 Animal Rights March on Washington scorned the use of horses to plow cornfields and the Amish who use them as an example of animal exploitation. When asked what might be done instead, the reply was a glib, "Get a tractor" (Schmorrow, personal communication, 1990). Of course, such advice fails to appreciate the fact that the use of fossil fuels in the tractor's construction and operation will, in the long run, do more exploitative damage to all life on Earth than plowing with horses ever could.

The present paper looks at both the world of science and of antivivisection, making the case, as did the Native American Chief Seattle in 1855 to the President of the United States (Ulrich, 1989b), that humankind's hunger is eating the Earth bare. What is happening to the Earth is happening to its children. We are all a part of the fabric, and what happens to one form of life will eventually happen to all, regardless of our position relating to animal rights and animal wrongs.

Where do we go from here?

The issues involved in the debate concerning the ethics of animal research are complex, requiring a good deal more wisdom and less emotion to lead us toward policies that will work out best in the long run for both people and other animals. It is imperative at the present historical juncture that scientists reevaluate animal research. A scientific reevaluation should consider the crisis situation that reflects the way in which human beings treat the Earth's environment, which we share with other animals.

Limits to growth

Trained as a behavioral scientist, I was taught to observe and respect data and report the truth as I see it (Ulrich, 1989a). In short, the conclusions from my years as an animal researcher are that Chief Seattle was correct: humankind's hunger is eating the Earth bare, and what is happening to it is happening to its children.

In 1970 an international team of researchers began a study of the implications of continued worldwide growth (Meadows, Meadows, Randers & Behrens, 1970). They examined five basic factors that determine and, by their interaction, ultimately limit growth on this planet. They were: population increase, agricultural production, non-renewable resource depletion, industrial output and pollution generation. The researchers fed data on these five factors into a global computer model and then tested the behavior of the model under several sets of assumptions to determine alternative patterns for humankind's future. The resulting message of that study is that the Earth's interlocking natural system of global resources cannot support the rate of depletion that is currently taking place. As humans continue their attempt to expand and grow, the results will be the same: resource depletion and the ultimate destruction of life-giving habitat will increasingly manifest itself all around us (Rifkin, 1980).

Wise people have spoken of the limits to growth for some time. A sense of self-discipline formed the very root of everyday life for Native Americans, who inhabited this land before the utopian "no limits to growth" ethic, imported by European settlers and based upon the Newtonian world-machine view, came to prevalence (Lopez, 1980).

Our mass-consumption, human habits that override the universal truth that "there are limits to growth," are perhaps too often a direct function of the widening gap between our actions as scientists and what we know as the truth. Research has shown that we must use fewer resources, but we continue to over-consume. The proposition that our human well-being requires ever more resources to pour into research institutions that often demonstrate allegiance to ideas like "Big is beautiful," "More is better" and "Money is our savior," rather than a loyalty to the ethics of true discovery, must be reconsidered. We, as scientists, should at least consider the possibility that science, as Anne Wilson Schaef (1987) suggests, has become as severely addicted to excess as the larger society.

We certainly should not overreact to criticism of animal research when it truly benefits the survival of all life on Earth. Indeed, many animal-rights activists are not the enemy of science they are perceived to be. What is needed is more good-faith dialogue between individuals who may, for the moment, consider themselves the "enemy" of the other although, in the long run, they may discover that, in fact, they are allies facing the same foe.

Many life forms, human beings included, seem prone to respond to short-term contingencies. If it feels good, do it again. This is especially true of non-human animals when confined and forced to exist in unnatural laboratory settings. It is, in fact, these same animal beings that have been removed from more healthful natural settings so that we might learn from them new lessons for physical and mental health. Humans, unfortunately, seem to be unbounded in their ability to overcome limits of resource depletion compared to other life forms who live in closer balance between the intake and output of energy and matter (Barry, 1977).

Undoubtedly our survival has been, to a point, seemingly blessed by our research in as much as the human species has proliferated in number (Ehrlich, 1971). Yet within that observation lies the very seeds of the animal rights-animal wrongs debate. Our search for better understanding has led us toward a behavioral dilemma in which we must face ourselves in the eyes of the very animal subjects that we keep under the conditions of extreme discomfort experienced by *all* beings in bondage.

Try as we might, scientists cannot escape the fact that if our animal subjects are reasonable models of our own most intricate actions, then they must also be respected as we would respect our own sensibilities. We cannot logically defend our scientific work with animals on the basis of the similarities between human beings and animals but defend it morally on the basis of differences (Ryder, 1978).

Certainly many people criticize animal research in what may seem to be irrational ways. It will not be wise, however, if the psychological society fails to note that increasing numbers of scientists as well as clinicians are criticizing animal research on scientific grounds.

No matter where we turn to place blame for violations of animal rights, a finger will always point back at our own individual behavior. For this reason, it is important to consider the conclusion made by psychologist Michael Giannelli (1985), who said, "I believe the most valuable things we have learned through animal experimentation are insights into the human mentality. These insights have arisen from direct analysis of researchers at work, not from tenuous extrapolations to ourselves based on animal behavior in highly artificial laboratory environments. We have learned that otherwise compassionate people can become remarkably desensitized and detached from the suffering they inflict upon animals. We have learned that highly intelligent people can be engaged in the most trivial or eccentric research, yet convince themselves that their work is important."

Have we perhaps, as Pogo (that little animal rascal) noted, met the enemy and it is us? Indeed, it may be. Increasing numbers of scientists and clinicians are criticizing animal research on scientific grounds, maintaining that newly-developed methodologies are less expensive and more valid than animal studies. Given that the prohibition of all animal research is not possible at the present time, the important questions concern whether our current level of support can be justified in the spirit of good science. If it is shown that over-reliance on animal research wastes billions of dollars and often yields misleading results, then our level of support for animal research must be reappraised. Animals, like humans, are a part of and also affect Earth's interlocking system of resources.

Historical impact of animal research

Some animal research proponents assert that all major medical advances have relied on animal experimentation (Stanford Committee on Ethics, 1988). However, others suggest that most of the key discoveries in several areas, including heart disease and cancer, were made through clinical research, observations of patients and human autopsies. Animal research in those areas served primarily to "prove" in animals what had already been demonstrated on people.

The scientific tradition that medical hypotheses must be "proven" in the lab has had unfortunate consequences in that effective therapies have been delayed because of the difficulty in finding an animal model that "works." For example, research with the animal model of polio resulted in a misunderstanding of the mechanism of infection. This delayed the development of the tissue culture, which was critical to the discovery of a vaccine (Paul, 1971).

Brecher (1963) points out that misleading animal tests can be devastating for human health. Prior to 1963, all 27 prospective and retrospective studies of human patients showed a strong association between cigarette smoking and lung cancer. However, almost all efforts to cause lung cancer in laboratory animals failed. In fact, a leading scientist wrote in 1957, "The failure of many investigators...to induce experimental cancers, except in a handful of cases, during fifty years of trying, casts serious doubt on the validity of the cigarette–lung cancer theory" (Northrup, 1957). This lack of correlation between human and animal data delayed health warnings for years; subsequently, thousands of people died of cancer who may not have, given proper warning. How could widespread beliefs about the value of animal research be so inaccurate?

It is my opinion that the problem frequently starts at the academic level. Many academic scientists are animal researchers, and those who do non-animal research are reluctant to offend the powerful animal research establishment. Academicians teach medical students and graduate students, write the textbooks students study and edit the journals that all professionals read. Thus, they have the ability to disseminate widely what sometimes turns out to be an inaccurate account of medical history.

This is no less true in my own department. Animal researchers totally dominate the Animal Care Committee and, in spite of observations by many (including the official university veterinarian) that our facilities are not up to federal code, animal caretakers seem more likely to be hired if they can be counted on to guard the status quo, as opposed to pressing for reform (Bradshaw, personal communication, 1990). Animal research ethics, like other behaviors, evolve according to natural laws. As new discoveries are made which question current practices, swift change should not be expected.

In May 1990, I submitted a straightforward, opinionated manuscript, based on my experiences as a lifelong animal researcher, to *Psychological Science* for consideration as a general article. It was entitled "Animal Research: A Psychological Ritual."

In June, Dr. Estes wrote what I considered a kindly and fair response, saying, "I don't doubt at all your expectation that the topic of your article will receive increasing discussion in coming years. I have to judge, however, that whatever part *Psychological Science* may play in this discussion, it cannot start with your present article. The decision is not an easy one, for I recognize that articles that make use of and support research with animals do appear in *Psychological Science* and it is reasonable that opposing views should be able to get some representation. However, your manuscript, though most interesting and thought-provoking, is too much a personal document to be suitable for a scientific journal. You could perhaps get a hearing in *Psychological Science* if you would prepare a *more soberly reasoned article."*

In July, David Johnson's "Animal Rights and Human Lives: Time for Scientists to Right the Balance" appeared in *Psychological Science* as that issue's *Psychology in Washington* column *(PS* 1(4), 1990, pp. 213-214). I read it and wondered: was the hearing provided for David Johnson's very personal document based on some standard of sobriety and reason somehow missing in my submission and in my mind? I thus wrote Dr. Estes again and asked him to reconsider his decision not to publish my manuscript in light of the space given to David Johnson's views on the animal rights issue.

I was concerned that the "turn down" by *Psychological Science* was an indication that our profession has come to a place where it is willing to allow as proper the expression of one side of an ethical debate while rejecting expression of an alternative perspective. My answer came in September in another personal communication from Dr. Estes.

> I appreciate the thoughtfulness of your last letter, and I think I'm probably as sensitive as most to the issue of fairness. Yet, so far as the manuscript you submitted is concerned, I can't even after considerable meditation find any way to improve upon my last communication about it.
>
> However, the context has changed since then in that the column on animal rights has evoked a flood of partisan reactions and I find myself and the Journal more involved in the animal rights issue than I had ever bargained for. I can appreciate Johnson's position well enough. He is constantly besieged with calls for assistance from behavioral scientists whose laboratories are literally under siege by animal rights activists. Having been led to think more about the problem, I think the issues involved are dreadfully complex and that a good deal more wisdom and less emotion are needed to lead us toward policies that will work out best in the long run for both people and other animals. Thus, referring now to the handwritten postscript to your letter, though I can't predict with assurance how I would react to an article I haven't yet seen. I certainly would give careful and sym-

pathetic consideration to a reply to Johnson of the kind you suggest you might try your hand at.

Based upon that letter, I decided to attempt to compose an article, somewhat as a reply to David Johnson. Even more so, however, I felt it was important to make an effort to move toward healing some of the wounds which have developed between attempts by science to better understand and direct human actions, and views held by antivivisectionists that our efforts have been unproductive and, at times, cruel. Above all, there remains the ethic to do better by our animal brothers and sisters.

Coffee with David Johnson

Certainly David Johnson is not wrong reporting his observations in the July issue of *Psychological Science*. Indeed, it is, as he notes, "time for scientists to right the balance," but it cannot be equilibrated by presenting only one side. There are, of course, excesses within the animal rights movement. There are also excesses within the animal research institutions which cause unnecessary cruelty. There are excesses within each of our own lives, regardless of how we align ourselves on the animal rights issue. No doubt most of us often eat too much, drink too much and drive too long in our gas-guzzling, exhaust-spewing automobiles. We all need to do more with less, whether we are scientists in a research laboratory or anti-fur activists at a fur fashion show.

I consider myself to be for animal rights as well as for animal welfare, and as a psychological researcher who ceased my practice of shocking animals, I frequently find myself approached from within the "animal rights" community. Like all groups, including "scientists," it is a community which sometimes deals harshly with those who not only do scientific research, but also those who wear leather or fur, who hunt or trap and who eat meat.

Regardless of our profession or what we consume to keep ourselves alive and clad, human beings are no doubt the most unabashed overconsuming wasters that have ever lived, and during the past 20 years our species has increased by 47%. We are fouling the surface of the planet as we burn the fuels to develop and transport special foods. We are destroying animals, birds, fish, insects, fresh water, air and earth. We seem to convert everything we touch into cities, suburbs, sewage, smog, roads, rust, and ever-enlarging fields upon which big tractors inefficiently roam to grow more things to eat. Meanwhile, at schools, churches, scientific and other conventions we insanely proclaim our enlightened sanity and ascendancy over other life forms; simultaneously we conduct research to prepare our young to use the deadly weaponry, as well as the medical technology, we design to counteract our unnatural fear of death.

Certainly many people greedily consume far too much animal flesh. Yet name-calling by "animal rightists," who indulge in persistent, self-righteous promotion of the exclusive killing and eating of plants, will not produce a solution to worldwide suffering of laboratory and other animals. There is no way to avoid the fact that life feeds on death.

Here again it is important to point out that the problems of animal use are those faced by all life. Greater temperance in our eating habits is one good step for us to take. Also, we should realize that we are all hunters engaged in stalking and stocking our food. Some people hunt with weapons in places where nature is more similar to times past. Others hunt through the ads on TV and in newspapers, and listen to voices over the radio telling them where to drive to bag the best bargain, and how one car with a certain type of tire, oil and gas is a better vehicle for stalking the elusive prey of health and happiness. Others, called scientists, hunt for new information within the actions of animals they call subjects. All human hunters must become more sensitive when pointing fingers at other human hunters.

Doing more with less

For me, the overriding research issue for scientists, as with all other human activities, relates to the question of how to solve our own overpopulation and the impossible continuance of the resource depletion it causes. What is the greatest ethical good for the totality of life on Earth, and how do we get ourselves to abide by it? Much animal research today is, in my opinion, simply an example of "bad science." Many students and their advisors, who don't know the difference between a male and a female bird, also have no idea nor real concern as to whether the data they are generating will ever benefit anything other than their own graduation or career advancement. Pigeons kept confined at 80% body weight in home cages that don't allow them ever to spread their wings, take a bath or relate socially to other birds provide questionable models for humans in a free operant setting. I realize, however, that giving up long-established habits is next to impossible, so if today you are a scientist still studying pain aggression (Ulrich and Azrin, 1962) by shocking four rats, try using two instead.

Let us all carefully analyze our own behavior. Why are we doing what we are doing, especially if it is being done to another living creature that we assume is like us (otherwise why use it as a model for the discovery of data to aid humans) and thus would not appreciate what is happening?

Abby, dear Abby

Ann Landers frequently addresses the animal rights issue, as she did in a column titled "Aren't people's rights greater than animals'?" This one she prefaced by saying, "My good friend Dr. Michael DeBakey, chancellor of Baylor College of Medicine in Houston and chairman of its department of surgery, has written a splendid article that appeared in the *Washington Post.*" Then, under the bold-type headline "Medicine Needs Those Animals," she reprinted parts of DeBakey's article.

However, Ann sets us up. She tells us that DeBakey is a "good friend," not a "bad friend," and his article, which appeared in the *Washington Post* (Washington: the name of our founding father and our capital and a lot of elementary schools) is "splendid." Ann knows a lot about managing human behavior; so does DeBakey; and so do we all as behavioral scientists. Thus we all must be especially careful *to allow discussion.* Scientists should deplore any unfair or exaggerated anti-animal rights propaganda with the same vigor that we deplore misinformation by animal rights activists.

Researchers who take a stand for animal rights sometimes discover that they are discriminated against. They are made to look silly, not respectable, a discredit to their profession, not to be taken seriously by a public frightened by images of activists as terrorists. The American Medical Association states in its Animal Research Action Plan of June 1989, "one has to...isolate the hardcore activists from the general public and shrink the size of the sympathizers.... The animal activist movement must be shown to be not only anti-science but also responsible for violent and illegal acts...and a threat to the public's freedom of choice (to eat meat, wear furs, attend rodeos, etc.)." I feel such rhetoric only makes things worse and clouds the fact that all of us, animal researchers as well as animal-rights activists, are addicted to a bloated, artificial lifestyle propped up by environmental exploitation.

As scientists we have all no doubt conducted research that has produced positive, life-enhancing effects. Resources that are used up to produce those effects, however, are often far too expensive in relation to the overall benefits they produce. The negative effects of the resource destruction required for much of our research, as presently conducted, may be found more damaging, in the long run, to all those suffering the effects of human sickness than will be the positive short-term gain made by our continuing use of other life forms in an exploitative fashion. Today the entire planet Earth is at risk. It is easy to envision the future as one of ever-increasing environmental degradation, poverty and hardship among ever-declining resources in an ever more polluted and populated world. It is indeed time for scientists to right the balance. A good place to start is within our own society and within our own daily lives.

References

Berry, W. (1977). *The Unsettling of America: Culture and Agriculture.* San Francisco: Sierra Club Books.

Bradshaw, L. (1990). Personal communication.

Brecher, R. (1963). *The Consumers Union Report on Smoking and the Public Interest.* Mount Vernon: Consumers Union.

Curtis, P. (1978). New Debate Over Experimentation with Animals. New York: *New York Times Magazine,* December 31, 1978.

Ehrlich, P.R. (1971). *The Population Bomb* (Rev. ed). New York: Ballantine Books.

Giannelli, M. (1985). Three Blind Mice, See How They Run. In M.W. Fox & L.D. Mickley (eds.) *Advances in Animal Welfare Science* 1985/86, Washington, D.C.: Humane Society of the United States, pp. 109-164.

Johnson, D. (1990). Animal Rights and Human Lives: Time for Scientists to Right the Balance, *Psychological Science, 1,* pp. 213-214.

Lopez, A. (1981). *Pagans in our Midst.* Rooseveltown, NY: *Akwesasne Notes,* pp. iv-xviii.

Meadows, D.H., Meadows, D.L., Randers, J., & Behrens, W.W. (1972). *The Limits to Growth: A Report for the Club of Rome's project on the predicament of mankind.* New York, NY: Universe Books.

Northrup, E. (1957). Men, Mice, and Smoking, *Science Looks at Smoking.* New York: Coward-McCann, p. 133.

Paul, J.R. (1971). *History of Poliomyelitis.* New Haven: Yale University Press.

Rifkin, J. (1980). *Entropy.* A Bantam New Age Book. New York: Viking Press.

Schaef, A.W. (1987). *When Society Becomes an Addict.* New York: Harper and Row.

Stanford Committee on Ethics (1988). Animal research at Stanford University: Principles, policies, and practices, *New England Journal of Medicine, 318,* pp. 1630-1632.

Ulrich, R.E. (1989a). *Observe and Respect: The Experiment of Life.* Littleton, MA: Copley Publishing Group.

Ulrich, R.E. (1989b). *Rites of Life: A book about the use and misuse of animals and earth.* Kalamazoo, MI: Life Giving Enterprises, Inc.

Ulrich, R.E. and Azrin, N.H. (1962). Reflexive fighting in response to aversive stimulation, *Journal of the Experimental Analysis of Behavior 5,* pp. 511-520.

PART II
POINTS OF VIEW

CHAPTER 3
Education

Without question, Roger's view of universities has been shaped by his many years on various university faculties. This chapter contains Roger's critiques of the corporate university. In a 2007 Viewpoint article reprinted from *The Kalamazoo Gazette*, Roger charges that universities are concerned far too much with profit and refers to them as the new "Alma Corporate Mater." He has serious doubts about the exclusive emphasis on a college degree, and explains his belief that "our physical survival is intermingled as one with the Earth, air, water and a spiritual connection to that which remains the mystery of a power often found beyond our control that cannot be won with money or a college degree." Ulrich's strongest critique is that the corporate-influenced university "dominates our conception of how life should be lived."

For additional articles on this subject, go to www.lakevillagehomestead.org/media

◂ *Kids visiting kids during one of many farm days, when children from schools in the Kalamazoo area come to see what a real farm is like.*

What Constitutes "Education"?

Letter to the editor of The Kalamazoo Gazette

April 5, 2010

Dear Editor,
*   The Board of Education for the Mattawan Consolidated Schools claims in a Gazette Viewpoint, "We are failing an entire generation of children's education." Not mentioned once is what constitutes "education"! Mentioned often is money, funding, dollars, property values, costs and, of course, taxes. The Viewpoint ends by saying, "We must fight for our kids; they are not old enough to do it for themselves."*
*   Today more than at any time in history, academic institutions from pre-school to post-grad are not geared toward allowing for experiencing the truth of life and gaining the wisdom that ensues from life experiences. We have lost the art of searching for and teaching the truth as it is and have replaced it with the ethic of scrambling for money so we adults can continue our over-consumption lifestyles, keep paying the obscene salaries doled out to corporate CEO's, winning coaches, college presidents, and, indeed, most of us who represent the approximately 5% of the world's population that yearly use up approximately 50% of the Earth's bounty and in the process have become clueless regarding how to raise healthy food.*

Perhaps it is time for us to look at how the Amish manage to peacefully "fight" for children via their own privately-funded schools ??? from an agriculture base that has not lost sight of the importance of raising healthy food rather than how to make more money and profit off of debt.

Roger Ulrich

Racism and Exploitation in Academe "Festering Out of Sight"

Reprinted from The Kalamazoo Gazette, *January 2, 1989*

In recent remarks at a conference on race relations, U.S. District Judge Richard Enslen implied that "racism is one of the most problematic ideologies of the modern world." A follow-up editorial by the *Gazette* reminded us that "racism too often festers out of sight."

Certainly it always seems more recognizable in others than ourselves. For example, our press rightly admonishes South Africa for its racist apartheid and the treatment of Nelson Mandela, yet seems to ignore its Americanized form as it applies to political prisoners in the United States such as Leonard Peltier and other Native Americans.

Racism is a historic by-product of the process of civilization and organizing human populations into cities. This signaled a change in knowledge about how to manipulate nature and mistakenly attempted to decrease dependence upon natural resources.

Cities and the universities which grew with them are centers of technology and government which, in order to provide for the survival of civilization, are extractive of nature and the countryside. That the countryside may be populated by a growing and increasingly impoverished minority who does not wish to be exploited is irrelevant to the exploiter.

It is important not to romanticize the purpose of either the ancient or the modern city university-government complex. *Disguised behind ju-*

diciary jargon, it exploits the natural countryside unmercifully and to the fullest extent allowed under the laws of physics. When the extractive process no longer works, and there is no longer sustenance provided by country people, that civilization will recede and fail

In the United States our five percent of the world's population yearly uses 40 percent of the world's resources. Currently, an increasing number of that five percent are being added to the ranks of the exploited, as modern slaves maintained in prisons, mental hospitals, on reservations and in other ghettos.

Slavery as a condition of race became popular following the "colonization" of the Americas and the penetration of Africa. To excuse the butchery of American Indians and the slavery of Blacks, Europeans looked to universities and the "educated" elite who provided an answer from Aristotle's *Politics*. Minorities of color became the man-beasts of Aristotelian philosophy. This bigoted rationalization fit well for those willing to ignore American Indian prophecy, which maintains that what you do will return unto you in this lifetime or the next.

The ancient world often forced conquered people into slavery. Today, it's produced through an exploitative system of education. Let us consider how.

In 1970 researchers at the Massachusetts Institute of Technology explored the implications of worldwide exploitation and growth. The resultant message was that Earth's natural system of global resources cannot support the current rate of depletion. As humans continue their attempt to expand and grow, the ultimate destruction of life-giving habitat will increasingly manifest itself.

Unfortunately, the over-consumptory habits that override the universal truth that "there are limits to growth" are a direct function of exploitative practices at universities. The myth that our well-being requires ever-increasing resources poured into city, university and other government institutions, which have consistently demonstrated their allegiance to the ethic of "Big is beautiful," is *"racism festering out of sight."*

Institutions of higher learning often measure success by the money they take in: begging funds from any source hopeful that growth is our savior or wants their name on a building or door.

Relative to many institutions, Western Michigan University is not a big league exploiter. Nevertheless, new buildings are constantly going up. All need to be heated, like those already standing, ignoring that the smoke from our existing heat source is spewing dangerous amounts of carbon ash.

Maybee Hall was recently destroyed to clear space for the growth of a new computer building, forgetting that it was computers who told of the fate that awaits if we fail to contain our appetite for growth. Construction of buildings for the business college has begun, in spite of the fact that buildings stand vacant all over a city in which citizens would be better served were the business faculty to move into them and prove they could run a business.

Soon an Honors College building will be added to the menagerie; although it would be more honorable to save the resources thus expended, and heed what the Honors College faculty teaches about the limits to growth.

Universities could learn a lot from Amish communities. Amish respect the land, avoid waste, remain aloof from higher education and don't construct unnecessary buildings. Amish farms double as churches, retirement villages, nursing homes and schools which have produced alumni whose record of social problems are infinitesimally small compared to graduates of modern educational institutions.

Our university-led culture models the exploitation of all minorities who refuse to buy into the "no limits to growth" trap, causing the world to be regulated by people taught in schools that show by their actions disrespect for Mother Earth's laws.

We all face our own exploitative actions. Think about it. Racists are exploiters; exploiters are racists! All fit the model.

One answer lies in academe, where a great attempt is needed to stop the exploitative raids upon existing stores of natural resources under the false pretense of progress.

Finally, remember that we serve as models for our children, who place their faith in us for guidance; and in the end, we are all graded in Mother Earth's course on ethics, whether our performance be in academe or elsewhere.

Education Should Nourish and Promote Value that Exists in Farming

Reprinted from The Kalamazoo Gazette, *June 9, 2009*

In his inaugural address, President Obama gave notice to the world that we face a new era of responsibility. To people of poor nations, he pledged to work to make farms flourish, clean water flow, provide nourishment for starved bodies and feed hungry minds. Nations enjoying relative plenty were told that actions would be taken against indifference to suffering and the consumption of the world's resources without regard to impact. In short, his words expressed support for efforts to bring about a more healthy Mother Earth.

In many ways President Obama articulates an ethic which nurtured my roots growing up within an Amish Mennonite "agri-culture," which spoke to a way of life that would serve to be re-explored as a viable, present-day home, farm and education model. Within its examples of living a life of peace, love, forgiveness and simple subsistence farming, there was an attitude nurtured that needs to be re-introduced throughout society, exemplifying the importance of reaching out as a caring community and a new era of responsibility that would show greater support for the production of healthy food.

Schools at all levels need to add to the reading, writing and arithmetic curriculum the most important, basic knowledge of all—how to raise healthy food and respect for those engaged in that process. The time has come to consider the two most important cabinet positions advising the president as to what actions are necessary for the successful conduct of all business: the secretaries of agriculture and education.

First, the Secretary of Agriculture guides the system that feeds us. For too long now, the agencies responsible for assisting farmers have been allowed to drift into a bureaucratic abyss, wherein the need of corporate agribusiness to make a profit has trumped the need for supporting programs of healthy food production. When there is a fair balance between the cost of production and a benefit to labor, we will once again find ourselves back on the pathway to economic sustainability.

The second most important cabinet post is the Secretary of Education, who guides the institutions that must teach and reinstate the importance of agriculture. We must promote, via the Secretary of Education, a group of educational leaders who understand that in order to assure that no child is left behind, education must be redefined to include wisdom that goes far beyond processing a piece of paper that denotes the holder has a high school diploma or a college degree. Longer school days with an oppressive load of homework that often interferes with home chores must be rethought. Also questioned should be the push for classes on Saturdays that then expand into school all summer long, finally exploding into suggestions by some that we make college mandatory for all. Forced schooling does not define an educational system, but rather imprisonment.

It would be good for all of our leaders to attend to *Omnivore's Dilemma* author Michael Pollan's open letter to the *New York Times* to "Farmer in Chief" Obama:

> "There is a gathering sense...that the industrial food system is broken. Markets for alternative kinds of food—organic, local, pasture-raised, and humane—are thriving...a political constituency for change is building...by way of the movement back to local food economies, and more sustainable farming. Nations that lose the ability to substantially feed themselves will find themselves as gravely compromised in their international dealings as nations that depend on foreign sources of oil. While there are alternatives to oil, there are no alternatives to food.... Changing the food culture must begin with our children, and it must begin in the schools."

Everybody has to eat. No living thing can survive unless it has the food to sustain it. It is in this spirit that we must join together as people of both poor and rich nations to work side by side to make our farms flourish and to school our children toward understanding and experiencing that necessity!

In Charles Frazier's gripping Civil War novel *Cold Mountain,* the story is told of the heroine Ada:

"She was filled with opinions on art, politics and literature…ready to argue the merits of her position…a fair command of French and Latin… and the ability to render a still life with accuracy…yet she wondered what actual talents could she claim when she found herself in possession of up to three hundred acres of land, a house, a barn, and animals…but no idea of what to do with them…. Now the (slaves) were gone and she discovered herself…frighteningly ill-prepared…living alone on a farm that her father had run rather more as an idea than a livelihood."

As President Obama noted, we face a new era of responsibility wherein we must make farms flourish and clean water flow to nourish starved bodies and feed hungry minds.

Reevaluate What Being Educated Entails

Reprinted from The Kalamazoo Gazette, *February 14, 2008*

Paul Ferrini's book *The Silence of the Heart* names fear as the root of stressful relations and that peace leaves our hearts when we think something negative might happen against our will.

Fear and confusion are being felt these days all around the world. At home we worry about jobs, energy bills, foreclosures, violence when young people shoot one another, schools less lavish than those in neighboring communities, failing academic achievement according to testing standards set by state and federal "No Child Left Behind" guidelines and that our children may not be getting a good education.

Almost daily, the *Kalamazoo Gazette* reminds us of the Kalamazoo Promise without anyone explaining, in any great verifiable depth, just exactly what is being pledged beyond the most shallow interpretation of a promise of a free college education.

Recently, the *Gazette* announced in its editorial column and again in another long opinion thinly disguised as a news story that "we are on our way to becoming the education city."

Education is tossed about as something all must obtain without it ever being seriously defined or equated with little other than a college degree. It has come to the point where our governor and state school officials are strongly suggesting that college become mandatory with the sole advantage being economics (the making of more money), which of course will not be assured for the students.

I wish to beg on bended knee that no one take my words to imply disrespect for educated people, a college degree or making a good living. Indeed, I feel, perhaps with a bit of uncalled-for arrogance, that my life represents all three.

I have spent my life among well-educated folks; first in my youth on the Amish-Mennonite farms of my family, friends and relatives and then, after many years working my way through colleges, I became a university professor with a doctorate, retired in 1998, whereupon I returned to farming and continue to make a good living.

Certainly no evidence exists to show that those with doctorates are more endowed with common sense than the non-college educated.

What is missing in all the talk surrounding education and the Kalamazoo Promise is any deep discussion of what is meant by "a person being educated," or, more importantly, who will feed those on their way to becoming educated beings both during and after the educational trip? Given the current course offerings at all levels of schooling, will there be anyone left who has acquired the skills necessary for producing the food our communities need for everyone's survival?

Does the Kalamazoo Promise promise to raise food?

Does it promise to serve us eggs and bacon or promise to raise the hens and hogs needed for food?

No! Our educational system, from top to bottom, desperately needs a revamping. The governmental agencies we expect to educate our children, that we assumed would show a lasting commitment to teach future generations how to care for the air, water, earth, and the agriculture upon which we all depend for food, is floundering.

In its wake we are left with a dysfunctional legacy that Joe Salatin aptly describes in his book *Everything I Want to Do is Illegal: War Stories from the Local Food Front.* Salatin shows how governmental guidelines have, over time, entrapped us into supporting both school and agricultural systems that are constantly intimidating citizens into giving up more tax dollars via tactics which go so far as to illegally cover up the truth of what is being done.

Sometimes, when I say these things, I am accused of being against government, education, college, teachers, children, not understanding the global economy and advocating that everybody should begin living like the Amish.

The truth is, however, that I deeply honor the art of teaching and a society that allows children the opportunity to experience and understand the truth by living it and having elders who do their best to model the ethics that serve younger people on their way to becoming mature adults.

If you were to read Salatin's book and another by Jo Robinson, *Pasture Perfect,* you will be the beneficiary of a promise that has as much integrity as any promise made in Kalamazoo.

The books also provide a perspective far different from that which is offered in the *Gazette* or in most courses taught in our schools.

If we truly wish to become the education city, we need to seriously re-evaluate what being educated entails. Only then can we become reconnected with the level of wisdom which, when modeled by adults, more honestly fulfills a universal promise that will truly leave no child behind.

Just What Does "The Kalamazoo Promise"?

June 2008

Just what does being educated
   Mean for you and me?

Does a promise made by Kalamazoo
   Get you educated free?

Does knowledge come for nothing
   With a college framed degree?

As we recall in days gone by
   Experience carried a fee.

Yet Kalamazoo keeps promising
   You can go to college free.

Where you're never taught to raise your food,
   But to think how smart you'll be.

So dare we ask if getting smart
   Changes "un" into "skilled"?

Does dressing up in suit and tie
   Make those who don't "unskilled"?

And does it really show we're wise
   Thinking wisdom comes unbilled?

And what's behind the money game
   Causing folks to leave the farm

To not produce but rather think
   Midst the myth of money's charm

That a promise made in Kalamazoo
   Can ward off hunger's harm?

And when we can't afford our rent
   Nor pay for gas or food,

And getting drugs for nothing
   Doesn't mesh with Pfizer's mood.

It is then we'll be reminded
   What the promise did not include.

And in the mirror of a later now
   Is this what we'll see,

A world of hungry people
   Who got "educated" free?

Never taught to raise their food,
   But instead how smart they'd be!

Some Food for Thought

Reprinted from The Kalamazoo Gazette, *December 15, 2005*

In the Dec. 7 *Kalamazoo Gazette,* George Erickcek of the W.E. Upjohn Institute said that Kalamazoo "is part of a trend that is moving away from producing for a living to thinking for a living." This got me to "thinking" about my own experience of retiring in 1998 from living as a "thinking teacher" and returning to living as a "producing farmer."

What I would like to know from Erickcek, Kalamazoo Mayor Hannah McKinney and the other economic experts who shared their "thoughts" in the Dec. 7 article is their "thinking" in regard to "thoughts" I've had as a function of reading about their "thinking" concerning their economic convictions of "thinking" versus "producing" for a living.

In short, what do they "think" is the best kind of farm on which one might raise the food they may be "thinking" of eating at their next meal? In my case, what type of feed should I "think" about growing to feed the chickens (from whom I am "thinking" of collecting eggs), and to feed the cattle, hogs, ducks, turkeys, etc., that I had previously been "thinking" about continuing to raise for folks to eat?

To keep up with current trends, maybe I should "think" more carefully before "unthinkingly" bungling along the path of food production for a living, as opposed to maybe trying to get rehired as a "thinking teacher." I would have The Kalamazoo Promise of ever more college students coming to be "taught" "thinking" and a supposedly endless supply of money to help me make a living.

Besides, any "thinking" person knows that "thinking" about spading the garden and mucking out the barn is a lot easier than "producing" the actual labor. Plus, as a student of Western Michigan University's Honors College once pointed out to me in all seriousness: "Raising food just wasn't that big of a deal, as you could always buy some at Meijer or just go to a restaurant." "Think about that one!"

Education Today Needs to Embrace, Foster Understanding of Agrarian Lifestyles

Reprinted from The Kalamazoo Gazette, *April 25, 2008*

USA *Today* and TDW and Associates CEO Tom Watkins, in a *Kalamazoo Gazette* Viewpoint, all support raising the mandatory age for compulsory school attendance to 18. Each mentions farming as a factor in their reasoning.

USA *Today* suggested that "in the nation's post-agrarian economy it makes sense for states still at 16 to raise their compulsory age." Watkins bases his reasoning for "raising the dropout-age" on his assumption that "it is a logical move in a knowledge-driven world," claiming that we have "moved from a society where you needed to lift for a living to a society where you now need to think for a living."

It is not true that we exist in a post-agrarian or post-lifting society. More realistically, we are in a post understanding of lifting and agrarian lifestyles. Post means past. To suggest that we are living in a world past the need for the agricultural economics that provide us life-giving food is misleading.

Agrarian refers to our relationship with the Earth and the thinking that encompasses the science, the art and the occupations concerned with what "lifts" our food from the Earth to our dinner table. If we don't eat, we will no longer lift or think.

Forcing all students to sit longer either in high school or college classrooms when some are convinced that they can better learn elsewhere how to live useful lives is not a good idea. Conventional schooling often seems to assume that students are naturally lazy, need to be forced to learn through incentives, kept busy with homework, and bullied into running the endless gauntlet of standardized tests. To demand by force of law that all students remain longer within such a system will mean reconfiguring our schools into prisons.

If we in Kalamazoo wish to become "the education community," we must seriously re-examine what being "educated" entails and elect school board members who will hire officials at all levels who have skills beyond being highly educated "panhandlers" good at the salesmanship necessary to get voters to support still another tax increase. Too often, most of the money raised from such a pitch goes into pockets other than those of the elementary level school teachers who deserve it the most.

Today we desperately need leaders who will foster novel solutions that lead to positive transformations within our schools and their associated communities. Schools must move toward doing a better job teaching a respect for community-supported agriculture beyond having children simply looking at a cow or a potato on a computer screen.

Further, such a transition must include more than learning about food production; it must encompass the totality of the "inconvenient truth" of Earth's current crisis. We need to go beyond talk and allow for hands-on experiences relating to the production of what we can healthfully eat as well as food storage, processing, distribution and its preparation—all of which relate to our community's intellectual, spiritual and healthy development.

Mini-gardens at schools and at cooperative agrarian sites must become funded platforms for creating improvements in the total school environment. As a first step we need the media to become less shackled to the education industry and seriously explore and report opinions beyond those commonly held by corporate America.

When young people finally find learning meaningful and thus useful, we can then hope for a re-connection to the level of enlightenment that honestly fulfills the promise of "no child left behind."

Using Amish as a Guide, Not All Education Requires Classrooms and Degrees

Reprinted from The Kalamazoo Gazette, *June 26, 2008*

Having read Diether Haenicke's column about the crisis in American education, I decided to share my perspective based on the roots of an education which stems from deep within the Amish-Mennonite culture.

The crisis in American education, I believe, is already upon us. The hunter-gatherer society and the later agricultural economy were both based more exclusively upon what was produced on the land.

Society has more recently evolved into a war economy. In this economy, our well-being is dependent upon the production and shooting of guns, which are used not only in the so-called "War on Terror," but also by terror-filled children who, in turn, bring terror to the streets and the schools of Detroit, Los Angeles and Kalamazoo. Haenicke notes that high school graduation rates for school districts serving the nation's 50 largest cities range from a 77.1 percent high to Detroit's 24.9 percent low.

Not mentioned is the fact that the high school graduation rate for the Amish is 0 percent. Not graduating from high school is seen by Haenicke as the major cause of this catastrophic national dilemma, and he suggests further that money is not the root cause of our crisis in so much "as federal and local governments have pumped hundreds of billions into improving public schools…with little return on investment."

Rather than it being lack of money, Haenicke feels that "social pathologies, festering historical wounds, cultural objections, and possibly other factors dominate this dismal picture." In the long-run, he concludes, "We will not be able to shoulder the heavy social and economic prices" that stem from this "vastly undereducated urban population."

I would suggest we consider another possibility. Not only high school dropouts, but our whole society is "undereducated." In short, we exist in a society that has lost its agricultural roots. We keep implying that "educated" equates to having a high school diploma or a college degree, as opposed to children having experiences which teach them to behave in ways necessary to the continuation of a functional human civilization. Certainly not everyone should be expected to become Amish. Nevertheless, educational leaders should look into what these non-higher educated people are doing to bypass many of the problems we "educated" folks keep having. The Amish pull off their "non-educated" trick in spite of neither registering nor voting. This, according to a recent *Gazette* editorial, makes them guilty of being "apathetic" and "not concerned with the education of our children."

A recent letter from a colleague implied that the "truth" behind the constant complaining about children not getting through our public high schools has less to do with the well being of our children than it does with the money that sustains the lifestyles of some of the so-called "educated" and some of the media bosses who pander their cause.

My colleague's letter read, in part: "Roger, I have come over the years closer to your ideas that a much simpler way of life with more respect for our environment is needed. I see enormous exploitation of our precious resources, boundless greed for ever more, and the widespread senseless and appalling greed in our economic and academic institutions. I also have become rather pessimistic about the likelihood of change. Most educators do not set an inspiring example for their students and are infected by the cancer of economic greed and social and environmental irresponsibility. I fear nothing short of real disaster will change our society. Sorry to sound so glum, but best wishes and many blessings anyway, and thanks for writing."

As long as we insist on exclusively equating being "educated" with having a college degree and paying lots of money only to those who have one, the crisis in education will remain. In this regard, urban sprawl will continue to force farmers off the land. Cruel evictions from homes, such as the plight featured under the April 30 *Gazette* headline "Eviction day arrives," will persist and devastation from worldwide natural disasters will rule the day and, in the end, determine which children will be left behind.

Universities Are Run Too Much like Businesses

Reprinted from The Kalamazoo Gazette, *2007*

A while back, Western Michigan University President Emeritus Diether Haenicke titled one of his opinion columns, "Let's not run universities like businesses."

I find myself agreeing while at the same time feeling that it's a "done deal." Universities are already run "like a business" and are no longer controlled by the people who work in them, but rather behave in accordance to mandates from "Boards of Control" that allow "academic freedom" only to the extent that the education system makes a profit and spreads the consumer lifestyle. Students are viewed as customers of our services. More than other institutions, the corporate-business-university-combine dominates our conception of how life should be lived. That information then becomes the university curriculum. An effective censorship thus results wherein all "progress" at the university solidifies into a unified machine whose business is that of fostering greater "profits" and "economic growth."

Locally, the transfer of the Baker Farm from a food-producing enterprise into a more "profitable" business is a case in point. Those willing to speak against the folly of destroying farm space by covering it with buildings when buildings already stand vacant and uncared for all over the city are invalidated as extremists against everything a good community needs. Less and less appreciated is the fact that people need food and that hogs, cattle, corn and wheat cannot be raised in cities.

Those at universities who speak and act effectively in defense of safe water, air and earth and ethics that measure things in terms other than solely making money are "fired," "retired" and not given tenure. More farmland is thus "sprawled" over, allowing even greater profits for businesses that avoid taxes via the free labor provided by faculty and students of non-profit university facilities. These facilities and labor, however, are not free, but rather paid for by the parents and all other citizens via local, state and federal taxes or tax breaks. Students in a sense pay for the privilege of becoming new age slaves who via the tuition they pay to cover the costs of college credit as Research and Teacher Assistants are doubly robbed, as they and their parents go into debt for the privilege of keeping corporate stocks paying high yields to businesses that sell them cars, housing, guns, cheap food, entertainment, alcohol and grief counseling. This modern, civilized custom of making our children borrow from and pay their elders to learn what is freely and lovingly taught as an ethical matter of simple survival sets "civilized" culture apart from native aboriginal people and virtually all others in the animal kingdom and is the hallmark of exploitation.

Adding "injury to insult," corporate business law demands that its managers act primarily in the economic interests of shareholders…and are thus legally obligated to ignore environmental and human welfare issues if those issues interfere with profitability. As educational institutions become increasingly controlled by corporations, educational leaders like all others have less and less choice but to follow a path that leads toward making money over all other considerations. University Public Relations Departments keep their jobs by feeding the media information mainly geared to convince the general public that his/her education is doing a good job for the future well-being of children. Somewhere down deep, students unconsciously began to feel that they, like society itself, have become "consume everything addicts," and so why not go out, buy beer and have another Lafayette Street party.

Unfortunately, business corporations, unlike humans, do not have to die, but only merge. The organic reality that characterizes real life, thus, is lacking within our most formidable educational systems that are run not by people, but rather a collection of abstract rules incapable of behavioral shame or remorse. It is a trend in which the traditions of academic and press freedom are translated into objectives enshrined into laws that are watched over by a justice system beholden to those with enough money to determine who gets elected. Soon just like "open space is to urban sprawl," everyone, including faculty, *Gazette* writers, students, as well as emeriti presidents and research professors, must at some level "sell out" in order to keep their "jobs" and/or retirement checks coming in. While singing eternal loyalty to the new "Alma Corporate Mater" and to the freedom to roam the world defining anyone who disagrees as evil, we search not for truth and telling it like it is, but instead for money while trying to hide that truth. In this regard, everyone should read *Lies My Teacher Told Me* by James Loewen.

All of us are educators of tomorrow's children, who like ourselves must live the truth of the ethic that our physical survival is intermingled as one with the earth, air, water and a spiritual connection to that which remains the mystery of a power beyond our control that cannot be won with money or a college degree.

CHAPTER 4
Lake Village Homestead

Lake Village Homestead is Roger's ongoing experiment in living well with less. In this chapter, Roger tells the story of the creation and development of the Lake Village community from idea to reality. Roger and Lake Village Homestead have repeatedly been featured in local, regional and national media and recently celebrated 40 years of community life. Also included here are some of the many interviews and media coverage Lake Village Homestead has received from its early '70s origins until today—portraits that chronicle well the ongoing educational contributions that this working farm has made to the greater Kalamazoo community and beyond.

For additional articles on this subject, go to www.lakevillagehomestead.org/media

◄ *Sometimes it just feels good to sit astride a horse. Many are the Kalamazoo kids who get to feel that feeling for the first time at Lake Village and get an education via the actual experience, as opposed to via words or a picture as they sit at a desk in school.*

IN ROGER'S WORDS

Interview with Roger Ulrich
by Hilke Khulmann, June 10, 1998

Q: If somebody who is totally unfamiliar with your community asked you, "So what's Lake Village all about?" what would you answer?

A: [pauses] It would be very difficult for me, from my perspective, to be able to encapsulate an answer to that question. So I'd probably say, "I have no way in the world to tell you what Lake Village is all about." [laughs]

Q: What did you find inspiring about Walden Two *and how has that changed?*

A: What I found inspiring about *Walden Two* is that I liked Skinner. I tend to personalize things, and I liked Skinner. At that time in history I liked the idea of exploring and researching, I was getting more into *that* mode from a clinical psychology mode. I saw many problems in our society. So what attracted me to *Walden Two* was its social conscience. That for better or for worse, we should be doing some things with children early, and things like that. And all of the researchers and all of the other people that I knew in the Skinnerian movement drew me to reading *Walden Two*. And then I got very much taken by the thing that he had Rogers say to him in the very first page of the book. "It's a job for research, but not the kind you do in a university. You have to experiment with your own life. Teaching is alright to turn people on, but in terms of finding things out, you've got to experiment, and experiment with your own life." So that was, and still is, very important to me.

Q: Would you say that the passage about having to experiment with your own life was probably what attracted you most to Walden Two?

A: Absolutely. [pause] There is so much theory in the book. It's a theory that has almost become a religion, like with Juan and all the guys down in Mexico and so forth. Which is alright, but that's data. And they did experiment with their own lives, and experiments: they show what they show. So what they're doing down there is beautiful, what they're doing at Twin Oaks is beautiful, and all the others. It's just what's happening. Again, there is no experiment other than a real situation, which comes kind of from Rolling Thunder. So that very much governs my outlook on a lot of things.

Q: When did you begin to be influenced by the medicine man Rolling Thunder and other spiritual experiences? And when did your Amish-Mennonite background begin to resurface in your thinking?

A: The Amish-Mennonite ethic was instilled in me in my mother's milk, and there were times when that got to be a bit oppressive, particularly when the Christianity side of it excluded other forms of spirituality. That very early on became a stumbling block for me, and so I started arguing with my Sunday school teacher. So there were a whole bunch of years of arguing with the Amish-Mennonite teachers and so forth. But you never escaped the love and the caring and the family that was around you, and even though people disagreed with you they still loved you and cared for you. [pauses] The question again was?

Q: When did these various spiritual influences resurface in your thinking?

A: They resurfaced again out here, after coming back to the land with a more scientific, non-spiritual approach to things. Out here, the land almost demanded of me that I start to look at that which is, the great mystery which is the Sioux Indian characterization of God, the Great Spirit and so forth. And so I was drawn into and became a part of the right place at the right time. And the spiritual side of, "We don't control the world, we are a part of it, look at it, and try to flow with it as comfortably as you can and do the right thing."

Q: How important do you think were your hands-on experiences with drugs in making you realize that behavior is less predictable than behaviorism had led you to think?

A: I'd say that for many Americans, LSD, taking the pill, was an instant enlightenment in some ways. I mean, for a moment there…[pauses] It's a typical American European way of doing things. We are not going to sit in a cave for thirty years. Boom! We need it now! And it brought an awareness to myself and others of the sort that was very, very important and did, for me, provide data that I followed, as opposed to sticking with the laboratory experiments of shocking animals and all that kind of stuff. That became silly.

Q: *Let's get back to* Walden Two *for a moment. As a novel, would you say it presents a completely positive utopia, or do you see potential dangers in what Skinner proposes?*

A: Again, life is made up of contradictions. Utopia is nowhere. So, perfection is in the eye of the beholder, depending on who is calling the shots. From my perspective, it's a book to assign students in introductory psychology, which they're not even all that interested in reading and find it boring upside of *Ishmael* and *The Message Down Under* and Rolling Thunder and a whole bunch of things. It's not even a particularly well-written novel. [pauses] That's how I see it now. When I read it the first time…[pauses]

Are there dangers in a system in which we're presumptuous enough to say that there is no such thing as freedom, but we're free enough to arrange controls around children, adults, everybody? Of course. It's very, very potentially the same kind of thing that keeps reoccurring over and over again. It's just so much simpler to make a decision like that and not have anybody argue with you. As you go down that path, there is a lot of travesty that can occur with someone getting lots of control. I think Skinner was sensitive to trying to set up a situation in which there was a lot of faith by a lot of people and so forth. [pauses] Are there dangers in *Walden Two*? There are so many more monumental dangers out there than what Skinner said in that book that it's probably fairly inconsequential.

Q: *What are the more monumental dangers?*

A: Oh, a quarter of a million new people being added to us every day. And the fact that we have become so addicted to the abstraction of money in the first world countries that we have forgotten what wealth really entails anymore. So we're hooked on secondary reinforcers, and we'll kill to get them. We're worried about children being shot in our schools and we're wondering why that's happening, not noting that they're watching television in which we're going over and bombing the crap out of Baghdad because they're supposed to have used gas on somebody. [pauses] The problems that are in the world today have the function of the human being evolving into a life form that reacts very quickly to reinforcers that follow their behavior. It has taken us off on a whole variety of tangents that are just very detrimental to the survival of all life forms. And we're proud of our "progress." *Walden Two*? You can't even find it in a bookstore anymore.

Q: Let's get back to Lake Village. Was there ever in the beginning or later on the idea that you would create an alternative society? Was it ever the plan that Lake Village would be something like a micro-society?

A: I think in the very beginning, after having read *Walden Two* and being somewhat enamored with the power of positive reinforcement and all that kind of stuff and the realization that things needed changing. There was the hope that we could do things that would bring about positive social changes. [laughs] Then we realized that we were our own problem. That there was not much hope in us saving the world or certainly *me* not saving it while I had all these problems myself. I had enough trouble modifying my own lifestyle and behaviors that defined at times a fairly sick person. In terms of just looking in the mirror. [laughs] It's like, "Oh, why should I be the one that helps solve the world's problems?" I mean, good lord, look at our own! So it didn't last very long. It was just sort of an initial flash in the pan, an arrogant bit of self-righteous thinking that flashes back every once in a while but hopefully doesn't last through the night.

Q: So would you say Walden Two *was only influential at Lake Village for the first few months of its existence?*

A: I don't know. I mean, I'm sure that on any given day, I have a self-righteous Let's-go-save-the-world-you-can-do-it thought. But it doesn't last very long. Hopefully, you get older and wiser.

Q: So in the very early days, when you were still thinking you could do something grand, how did you envision Lake Village to be, in terms of membership, size, structure, child-rearing?

A: I didn't tend to do that much. I had done acid by the time I came out here and I'd had a lot of non-ordinary reality-type exposures. I had already gotten in touch with Carlos Castañeda, with a number of indigenous folks, so it was very much, remember to be here now. So again, what do you mean? This place is barely able to make it through the week, it's going to be a model? Why should Twin Oaks and Los Horcones think that they had anything whatsoever new to offer to a world that had the Amish Anabaptist movement and all the other movements? I mean there is nothing new under the sun. That's just—those are theses. Those are just what Harvard and Stanford and Michigan and all these places force people into. "Ah, we got to write something down, so we write stuff down." It has very little to do with…[pauses] It's trivial pursuit.

Q: If you are fairly sure that at least for you, Lake Village was not a grand plan to save the Earth, what do you think attracted other people to Lake Village?

A: It varies pretty much with everybody who came. Someone trying to get out of their home because they're scared that they are going to get busted for pot, or someone else is just out of prison, somebody got a divorce and needed a place to live. Each person came here for such differing reasons, usually because it was a cheaper place to crash than a lot of others and they didn't maybe have to clean up as much. You have 251 different reasons for why people came out here, you have to talk to each one of them.

Q: So largely it was personal, practical reasons that made people come out here? Do you think there were many people with utopian ideas, in the sense that they thought about how society should be and tried to actually do that at Lake Village?

A: While people were high or low or going through a divorce or whatever, they also had ethics and had feelings about how life should be and they had a spirituality about them, in varying degrees. Everybody deserves respect and everybody that came here had concerns about life forms in varying degrees, their fellow human beings, the animals. So, yes, everybody was also utopian. A utopian thinker. Some people win Pulitzer prizes writing about utopia and making movies, and others carry water and chop wood.

Q: Did you ever have people who came to Lake Village because they thought you were "doing" Walden Two? Was your connection to Walden Two anything that was important to the people who came, or that they even knew about?

A: Very little. [pauses] Most of the people who had any incentive of that sort were in the original group that came out. They were part of the Learning Village, from 1965 to 1970, and they were part of the Racine Conference. So early on, maybe our first six months or so. And then very quickly the reality that was really in *Walden Two*—"You've got to experiment with your own life!"—took over and we went in a hundred and one different directions. And I was not theologically imbued with, "It's got to stay a particular way." [laughs] I didn't support that, and the other elders that were around supported it even less. My wife Carole didn't give a damn whether it was *Walden Two,* she wanted a good goat herd. Skinnerians were always putting down abstract theories anyway, yet that's basically what Skinner put down in *Walden Two.*

Q: *Why do you think Skinner was never interested in living in a community? He has said himself that it was because his wife wouldn't like it and that being a professor at Harvard was just too reinforcing. Why do you think he was really not interested in living communally?*

A: His wife wouldn't like it and Harvard was very reinforcing, and going to conferences and talking about it was much more reinforcing. It's a pain in the ass to go and do what the kids at Twin Oaks did and what Los Horcones did. It's work. It's the same reason that early on in my life I said, "Well, I think I'll go to college instead of putting the hay up." It's a lot of work, so Skinner was just being pretty honest. Life is work.

Q: *Why did Skinner refer* Walden Two *enthusiasts to Twin Oaks rather than Los Horcones or Lake Village?*

A: Kat Kinkade visited with Skinnerians and it was very much a case where it was going to be a *Walden Two* experiment. Skinner did early on send people to us. I don't know whether it appeared in any of his writings or not, but I know that he did and he came out here and talked to people and so forth. But then, after a while, we didn't reinforce that behavior in him. I mean, they were coming back and sort of saying, "Roger isn't a believer!" [laughs] So it wasn't reinforcing with Skinner. Because Skinner became a convert to his own religion, and he started to believe what he wrote.

Q: Was Walden Two *ever talked about at Lake Village? Was it anything that you ever went back to and said, "Well, what are the suggestions in* Walden Two?" *Was it ever in your mind at all that that was the original impulse?*

A: It was in my mind when somebody was asking me questions like you are asking, and when interviews occurred and so forth. On the day-to-day basis this is irrelevant. For the most part, literally irrelevant. [laughs] We were much more benefited by things like having the people from my Amish church coming up here and helping us build a barn and telling us what they knew and stuff like that. And they *knew* something. It wasn't a bunch of, in a sense, literary bullshit. *Walden Two* was essentially irrelevant to our system, other than some of the getting us started, getting us thinking about it.

Q: So do you think it would be accurate to say that for Lake Village, Walden Two *was the impulse but also the point of instant departure?*

A: I wouldn't even say *Walden Two* was the impulse. I would say that Vietnam, the civil rights movement, my brothers on the north side of town getting busted on the head and the kinds of things that we were going through in the early '60s. That was much more important than *Walden Two* in why we came here. We came out here as social activists, no kidding about it. I mean, we weren't just playing around.

Q: So do you think it probably wouldn't make much sense to look at your child-rearing or decision-making and compare that to Walden Two *because that wasn't the plan anyway?*

A: Certainly we used reinforcing strategies, and used a lot of behavior mod strategies early on, with my children Tom and Tracy and the other kids. We ran the Kalamazoo Learning Village, we have films that we made where we're using tokens and are charting and all of that. But if you watched over the years, that became less and less critical, because we were charting the things we were getting done and not noticing that we were listening to what people told us, and what they were saying they were doing they weren't doing anyhow.

Q: Did you use behavior modification mainly with the children?

A: In a sense, we're doing it right now. The way I look and the smile, back and forth, we are in a control-counter control situation where I modify what you hand over to me and then I speak and that goes on all the time. Are we using b mod? In a purist sense, you cannot help but use behavior modification strategies.

Q: Let me modify my question, then. Were you consciously and in a planned way using behavior modification at Lake Village with the explicit intent of having planned, conscious, scientific social interactions?

A: Okay, again, back to earlier in the interview, when we were talking about when we came out here. I had already been through the Learning Village experience. We came out here, it was about the same time that I damn near killed myself shooting up cocaine and dropping acid. So a whole bunch of things were modified as a function of the fact that…[pauses] Some of the textbook things that said this is how you should run things didn't matter anymore. We were under the umbrella of a lot larger power than scientists wanted to acknowledge. There is a power around us that is the Great Spirit, is the part of Mother Earth telling us how to live, and very early on I acknowledged that consciously or unconsciously and started to go along with that. I was not religious about, "We're going to follow what we were talking about in the psych department."

Q: Why was the pre-school called the Learning Village? Why a village?

A: That was just a name. Eve Segal came up with the idea of a living and learning village. And we thought that sounded like a good name, so we called that the Learning Village and when we came out here we called it Lake Village. No rhyme or reason. You'd be surprised how little thought went into everything that's around you. [laughs] Maybe not, if you visit with us longer.

Q: How would you describe the role that you have played at Lake Village over the years?

A: Everything from a confused junkie to an absolute dictator. My role as I see it—again, in an idealized form—is to be a child of God, a child of nature, an ethical being, and to do the right thing as Spike Lee says in his movie and so forth. And my more perfect description of where I'm coming from is: I want to get rid of poverty, I want to see to it that we're not polluting the environment, all of these things. There's the other side of me that jumps in my van, gases it up, takes gas and oil from Kuwait, adding to part of the world's problems. I still strive to get my reinforcers, like you and like anybody else. How I'm behaving on a particular day varies from a spoiled child having a tantrum to someone who looks like a calmer elder who's 66.

Q: Kat Kinkade once theorized that if a group has a fairly strong person in their midst, it is unlikely that there will be another strong person within that same group, because there is a high likelihood that those two people would clash. At Lake Village, have you seen many people coming through who you experienced as your equals?

A: I would disagree with Kat in terms of what she defines as strength. It's not clear that always getting one's way or seemingly getting one's way defines strength. That might define fear, cowardice, not allowing somebody else to have a shot at running the show.

Q: Okay, so I'll modify the word "strength." How about power? A person who holds a lot of power in a community, for whatever reasons, and regardless of whether that's good or bad, the person is wise or foolish. How does that affect the community, and how does it affect who is attracted to the community?

A: Well, we need to get down to power again, and a lot of times that is defined by who has a lot of money. There are a lot of people in this country who have a lot more bucks than I do. Whether they have more power than I do, I have no idea. And as far as bucks go, maybe I have more bucks than Rolling Thunder, but I sure wouldn't equate that with power. So what do you mean by power?

Q: I mean power within a certain group to move that group in a certain direction. Being influential in the decision-making process. And being more influential overall than the other people.

A: Well, I have always been attracted to people who have a strong will, their own ideas, who kind of like to go and do their own thing, even though I get into arguments with them. Those are the ones I'm attracted to.

Q: Have those same people been attracted to living at Lake Village?

A: A lot of those same people are just like me. We keep our distance from one another. You need a certain amount of space. So there are ways in which you try to create an environment where it's okay to get your own way. And I like that.

Q: *But doesn't that point in Kat's direction? That someone who has very strong ideas about what they want is likely to have a radius of not having another person with strong ideas around them?*

A: Could be, could be.

Q: *Would you say that Lake Village strove to be democratic in its communal days?*

A: Well, again, I'm really not hung up on words and abstractions. Democracy? What's that supposed to be? We use that term free society as we go around the world crushing certain forms of freedom. [pauses] We follow to a certain degree the laws of Pavilion township, of Kalamazoo county, state of Michigan and so forth. And when I go out that driveway, I stop at stoplights sometimes. So, if you call that a democracy, then the answer is yes. We never tried to fool ourselves that we are somehow outside the boundaries of man-made laws, human-made laws or natural laws.

Q: *I didn't actually mean how you relate to the society around you, but rather internally. While interacting with each other, did you use democratic methods, like votes? Or consensus? What did you think of the idea of "professional government," Kat's term for the* Walden Two *idea of placing decision-making in the hands of those who know most about it?*

A: But *Kat* was defining who knows most about it.

Q: *But between those two poles, consensus and professional government, where would you say that you stand in your thinking?*

A: Probably in the middle. Because those are words, and it depends on what issue is being discussed. There's a whole bunch of things that I don't even want to be part of making a decision on. You know, where to put the pins in a boy's diaper. And other things are very important to me. And if something becomes very important to me, I can become a mean ass, I suppose, to some people. Not that I mean to be. It's just that I have my beliefs in certain things that I'd rather see happening.

Q: *And how were those issues resolved that were personally important to you?*

A: Well, a lot of them went unresolved. And others…[pauses] Some days I'd get my way, some days I wouldn't. Some days someone else would get their way, and some days they wouldn't. We bent, I thought, way over backwards to allow some people who I admired for their anarchy. Up to a point. But I was being somewhat anarchistic myself.

Q: *Do you think people who drift in and out of communities tend to have a slight problem with taking on responsibility?*

A: I think that's a growing Euro-American problem, particularly in this country. Not feeling responsible for your actions. And it's not unrelated to determinism and some of the things that Skinner was talking about. There is a reason for everything, so that we can always look and say, "Well, it's not *me."* [laughs] I wasn't breast-fed long enough, or this, that or the other thing. Yes. A lot of people simply turn and point their finger at somebody else and say, "It wasn't my fault." And certainly a lot of people that drift in and out of communities don't have to worry about the taxes being paid. A lot of the realities that keep a place like this going: they don't know they're there! [pauses] You've got a good point there. Responsibility is something that a lot of people skirt. In my opinion. And it takes one to know one. I sure skirted a lot of responsibilities in my life. [laughs]

Q: *What is the legal status of Lake Village? And what exactly was the financial arrangement? How did new people join, and what happened to their money?*

A: Okay. It started out that there already was a corporate by-laws in place, and we were a 501(c)(3) corporation running the Learning Village. And when the land became available, we purchased this land under the umbrella of that corporation. So we were legally bound by the laws of the United States, the state of Michigan, and so forth, to follow the rules set up for corporate entities, which means you needed a board of directors, members and so forth. In less legalistic times, it's been a tribe. So that group was the one legally responsible for the land. So that answers that part of the question. And that's still how it works today.

Q: How about new members' money?

A: Nothing happened whatsoever to new members' money. You are still your own person. You cut a deal, like, here's a spot to stay and in return for it you wash some windows. And we'll see where things go tomorrow.

Q: Did people pay rent from the beginning?

A: From the beginning, people insisted on calling it rent. From the beginning we said, no, everybody here puts in a specified share, and share had a real meaning. It was trying to say, "You're a part of it, you're sharing in the thing and you're not paying rent for another entity." In our culture, that was almost impossible to get across, and it still is almost impossible to get across. Because that is not the way we've been trained to think from the time that we were born. We weren't born in Indian tribes.

Q: So members paid a monthly share to the corporation?

A: Yes.

Q: Was there a defined amount of time that people were supposed to work, and could they either pay more money or work?

A: That was determined by where you lived. So if you lived in this place or that place, we had prices attached to how much space you had. And over the years, those kinds of things varied, depending on the economy of the greater Kalamazoo area.

Q: So not everybody paid the same share.

A: No, not everybody paid the same share.

Q: Who actually worked on the farm, and how was that organized?

A: Those expectations varied day to day based on one's ability that day to do things. Carole is pretty much on the farm all the time. So, she has a whole bunch of things that she does to see to it that the farm runs. Tony is here most of the time, he has a whole bunch of things that over six or seven years he has learned to do.

Q: *It's fairly clear to me how it works now, but I'm wondering how things were done in the early days. Who ran the farm?*

A: In the earlier days you were required to do even less. [pauses] Also, in the early days we had Work Sundays. So in addition to the share, you worked for two hours every Sunday.

Q: *But who performed the many daily tasks required on a farm?*

A: You know, it's almost hard for me to remember, given all the years, who was here during what time doing things. Because people came and went. For a long time Leon was here and pretty much did what Tony is doing now.

Q: *I'm still not clear on that point. How did you keep track of where people worked and for how long? Did you have a labor manager or something like that?*

A: Basically, I can understand where you're coming from. Because again, what happened to us, we became more like an Amish-Mennonite farm, where none of that stuff is specified. People got up at five in the morning and knew what they had to do and they went and did stuff. And over the years we have moved further along that path. That's probably difficult to understand.

Q: *Okay. Here's a different question. What do you know about Matt Israel?*

A: Matt Israel [pauses] was very, very strongly motivated, highly motivated by *Walden Two*. He was Skinner's student at Harvard, he was always talking positive reinforcement, and he was wanting to get *Walden Two* started. And in the Boston area he, a number of times, got groups together in apartment buildings and started mini Walden Twos.

Q: *What happened then?*

A: Oh, Matt, pretty much after the Racine conference went back out to the Boston area and then he, out of his interest in *Walden Two,* and funding it and so forth, got involved with group care homes for autistic kids. And then over the years his [laughs] his positive reinforcement thing went so that they were slapping kids around and really using heavy aversives to get autistic kids to behave. And he got a lot of notoriety and was having a lot of success—you know, this is off the top of my head—in that area, and then started to run into civil rights groups and people sort of saying you shouldn't be hitting...you shouldn't be treating...you shouldn't be using aversive control like that. So Matt, who was closely identified with and was always talking positive reinforcement, ended up heavily using aversives, cause it kind of worked.

Beyond Walden Two: Living the Experiment

Parts 1 and 2 of an interview with Roger Ulrich on the Lake Village Experimental Community, by J.Z.

Part 1

JZ: Dr. Ulrich, our local Psi Chi is interested in learning more about the community of which you have been a part for a number of years now and I'd like to begin with you telling us how it started.

RU: Thank you, Jenny. We are grateful for your interest. Lake Village grew out of ideas described by B.F. Skinner in his novel *Walden Two* and a program for preschool children which initially began in the early '60s at Illinois Wesleyan University. The research in early childhood was continued in 1965 when I came to Western Michigan University as head of the psychology department, and in 1967 The Behavior Development corporation was founded which became the legal umbrella for what became The Learning Village (which continues to date teaching preschool children in Kalamazoo) and in 1971 the Lake Village Experimental Community.

JZ: What were some of the original goals of the community experiment?

RU: Initially, as proclaimed in our non-profit articles of incorporation, they had to do with issues important at that time to experimental and applied behavior analysis…"promote research, training, education, and other applied practices relating to the education development and modification of human behavior".

JZ: How successful do you feel you've been realizing your goals?

RU: Well, at first we felt that we were not doing too well because quite often what was happening didn't fit the expectations for which our academic background had occasioned. We were, as Dick Malott calls us, "idealists" or, if you will, utopian thinkers.

JZ: What, for example, had your education led you to expect?

RU: I guess greater predictability. Neither the animals nor humans behaved as predictably as the textbooks led us to think. In *Walden Two*, Skinner had suggested that finding out how to do things the right way "was a job for research but not the kind you can do in a university or in a laboratory anywhere… you've got to instead experiment with your own life! Not just sit back in an ivory tower somewhere as if your own life weren't all mixed up in it." Yet in the novel *Walden Two,* he had everything working out perfectly according to his operant theories. Our research style, however, was not that of testing theories as much as simply following the data. Thus, when things didn't work out at Lake Village according to our expectations expressed in Skinner's novel, we didn't get discouraged but rather just accepted and used it as data; not that *Walden Two* was wrong, but rather that we were just operating under different conditions. We were in fact behaviorists…doing things and observing what we did.

JZ: And where did that lead?

RU: More than anything I think it led us toward a closer appreciation for natural law and nature.

JZ: For example?

RU: We spent more time with Amish people and Native American spiritual leaders, who taught us to appreciate what Skinner had said about the importance of experimenting with our own lives. Our Native American teachers kept insisting that in order to understand the truth you have to live it...not just read about it...and that there is no experiment other than a real situation. They taught us that the Earth is a living organism which just like us has the wish to feel well and be healthy.

JZ: And what about the Amish?

RU: I came from an Amish Mennonite ethnic background and grew out of an agricultural ethic which emphasized always trying to live well using fewer resources. To them small is beautiful.

JZ: So what effect did this Amish influence have on the direction taken by Lake Village?

RU: Good question! The Amish aren't especially impressed with higher education and a lot of talk, so I guess they reinforced my behavioristic tendencies to tend to behavior and follow the data. We thus paid less attention to academic issues (which far too often are less important anyway than a game of trivial pursuit) and instead dealt with real life in a more natural environment.

JZ: Can you be more specific? What do you mean by "real life"?

RU: Clean air, pure water, an unpolluted Earth on which to grow healthy food for healthy humans and animals. We became more focused on exploring what it takes to live a more sustainable environment ethic. The land upon which we live includes a mile and a half of Long Lakes North east shoreline. Approximately 115 acres of forest, meadows and wetlands, located ½ mile east of Portage, Michigan, city limits in Pavilion Township...so we are close to an urban environment and understand the pressures the Earth is experiencing from the human population explosion, which adds a quarter of a million people to the Earth daily. We came to appreciate the fact that we were a part of 5% of the world's population which annually uses 40% of the resources expended.

JZ: So what effect did that have on your utopian thinking?

RU: We realized that we had to change our own thinking. We had to embrace the first and second law of thermodynamics. You cannot create nor destroy matter and energy, but most importantly when you use a resource you forever render it into a less usable state. We profoundly met the exploiter and it was us. We realized that we could not be self-sustaining.

JZ: So how did you in fact actually live?

RU: Over the past 24 years there have been 251 people who have lived as members of the Lake Village family, with some working on the farm while others held jobs in the greater Kalamazoo area as teachers, social workers, secretaries, plumbers, roofers, musicians, students, etc., etc. Many members who originally lived at the commune later purchased adjoining land so that in addition to Lake Village's 115 acres, the community grew to more than 200 acres including the land adjacent to Lake Village with nine permitted homes housing about 30 people.

JZ: Why do you say permitted?

RU: When we first came to the farm, we went to the township and indicated that we wanted to start an experimental community. They explained that since we were zoned according to existing ordinances as an A-1 rural-agricultural district which permitted only single families, we would have to apply for permission to start a subdivision with sewers, curbed streets, street lights, etc. That wasn't what we had in mind, so we went in a different direction.

JZ: And this different direction was?

RU: We decided to stay rural…be an extended family farm (Amish-like) and grew from that. We pulled permits to remodel, insulate, put in bathrooms, a kitchen, and add additions to sheds, garages, barns, etc., and before long had people living all over the place, including in teepees, tents, campers, cars, all depending how cold it was. Our interpretation was that this was an ethical and morally-sound way to reduce, reuse and recycle resources and always tried to continue to upgrade for comfort and safety wherever it was that people were being sheltered.

JZ: What did your neighbors think?

RU: At first the going was rough, but as time went by more of our neighbors indeed turned out to be people whose first introduction to Pavilion Township was via living at Lake Village.

JZ: What were some of the incidents that made you feel that "at first the going was rough"?

RU: In the early '70s, not everyone was into saving the environment, civil rights, long hair and dope. At least to the same degree we at first were, so this caused problems that made the going rough. Eventually they worked themselves out as saving the environment and civil rights issues became more popular to support...more of our original detractors grew long hair as former long hairdo villagers cut theirs. Most importantly, we at Lake Village became convinced that living healthful lives did not include supporting the use in any way of either legal or illegal addictive substances...so we sincerely strove toward becoming a drug-free environment and have been closely involved with drug rehabilitation projects in cooperation with institutions like WMU, AA and NA.

JZ: What other, if any, involvement do you have with institutions outside of the experimental community?

RU: Each year hundreds of folks from local colleges, preschools, grammar schools, high schools, city recreation programs, as well as other interested people from all walks of life, visit the farm in person. Also numerous articles in newspapers, magazines, books and a variety of TV programs have chronicled the Lake Village story. In fact, just recently we received a call out of New York from Prime Time Live to do a story about Lake Village in relation to *Walden Two*.

JZ: What would you suppose the theme would be to that story?

RU: I don't know. You're never sure about how any story will be told...a lot depends upon the teller.

JZ: How would you like them to tell it?

RU: Actually, I wouldn't like them to tell it at all...we are the only ones who can tell it and then not very well. Besides, I'm not sure any kind of national publicity would benefit our community right now. I would rather tell it with you and like I did at the recent psychology colloquium; I'd tell some history, like I've been doing in the class Toward Experimental Living, I'd show the TV tape of Skinner and myself on the Dick Cavett program back in 1969, when Skinner was asked what was his most important work. He said *"Walden Two… it's all in Walden Two…that's the world of the future."* I think *Walden Two* was our base of departure and I would recommend reading it. The other book I would suggest in order for someone to get a feel for our relationship to, and our respect for, nature would be *Rolling Thunder* by Doug Boyd. Next, I think that anyone serious in regard to learning about the Lake Village Experiment should come out and visit.

Part 2
Learning to Expect the Unexpected

One word that can accurately describe the atmosphere of Roger Ulrich's home is comfortable. Pictures line the walls of the living room. Hung among them are various paintings and pieces of Native American art. One wall is entirely made up of glass panels, through which is a gorgeous view of Long Lake. A row of homemade spices, fruits and vegetables lines the wall also, under which is a homemade rock garden filled with plants, porcelain figures and cat food dishes. An old fashioned wood heater keeps the rooms cozy on this rainy Tuesday afternoon. Roger leans back in his chair and gently smiles, and when he does his entire face lights up and his blue eyes sparkle.

"The best way," Roger says, "to learn about the farm is to experience it." This view sounds very much like what he writes in his syllabus for the psychology class that he instructs at Western Michigan University, Toward Experimental Living. He writes: "There is no experiment other than the real situation, thus you must always expect the unexpected." Recently, Roger perhaps learned that lesson better than anyone. Over the past 20-plus years, there have been approximately 250 different people who have called Lake Village home at one time or another. However, at present there are just five people living there, all in a single structure. Not because the experiment was deemed a failure, not because no one got along or because no one wants to live there, and especially not because of a lack of living space. There are only five people out there, including Roger and his wife Carole, because Pavilion Township has nailed pink slips to almost every structure on the property claiming that they are unsafe as permanent living spaces for humans.

While Roger changes clothes to prepare for the tour of the farm, Carole chats about the weather, the farm and their children. Roger comes out in a full tan work suit complete with gloves and a pair of well-worn working boots. He starts off walking slowly down along the dirt road that has turned to mud almost as soon as the rain began, and points out with obvious pride various aspects of the farm. To the right is a huge fenced-in area where cattle are calmly grazing, obviously undaunted by the rain. On the left, however, the goat pen remains empty. Each and every goat has taken shelter in the barn. In front of him is a huge structure that closely resembles what used to be an old barn. The basic form of the barn can be seen if you look close enough, but through the years walls have been knocked down and rooms have been added as was needed. Despite all of the transformations that have taken place over the years, it is easy to see that nature was altered as little as possible and great care was taken in using recycled building materials when it could be done. As Roger makes his way through the buildings from room to room, he points out where people have lived and how the rooms, now barren, had looked when they were filled with people, furnishings and loving touches. Walking through the building becomes somewhat like walking through a giant tree house or fun house. There is a garage, a kitchen, several one-room apartments, offices, a basketball court, a community room, bathrooms, storage rooms, a recycling room with bins lining the walls and much more. The constant renovating that had been done through the years makes it quite easy to actually get lost in the structures. Several uneven stairways occupy the building, as do doors of various types and sizes. In one place a giant rock has been made into a step by building the structure up around it, accommodating it instead of hauling it off. The atmosphere of the community room is much like the living room in the house that Roger and Carole live in. Paintings, photographs and various other forms of art cover the walls. Some of the pictures include: B.F. Skinner's visit to Lake Village, a cousin of political prisoner Leonard Peltier, Rolling Thunder, who is a Native American medicine man and a good friend of Roger's. Also in pictures are former members of the commune, friends, guests and fellow WMU professors.

Roger opens yet another door and leads the way up yet another hidden staircase. When he reaches the top, the hallway goes off to the right and brass hooks line the wall for jackets. Halfway down the hall on the right is a doorway into a nice-sized bedroom with a closet, and a window with a great view of the land. Further down the hall on the left is a bathroom with a floor made up entirely of smooth stones. Roger explains that his daughter handcrafted it herself. At the end of the hall the apartment opens up into a living room with a wood heater exactly like the one in Roger's living room.

The kitchen is to the left and has a nice-sized island that separates it from the living room. On the far end of the living room, a glass door wall opens up onto a deck which is actually the roof of a structure below. The view, like many others from the property, is spectacular. It looks out upon the different animals and upon the lake. Roger sighs and sadly claims that this is where he feels someone should be allowed to live. The idea that someone judges this beautiful apartment unsafe appears to stir up many emotions in Roger. "To many people," he says, slowly shaking his head, "living here would be almost like living in a palace compared to their homes now, and to many of the people who have no place to sleep at night and no place to call home, it would be a palace." Roger shakes his head again and suddenly appears much older than he did before. As he continues the tour through the buildings, he leads up a short set of five or six stairs into an office. Instead of pointing out different aspects of the room and moving on as he has been doing, he sinks into a chair, looking very tired and worn. He begins to tell about how the recent conditions came about. A disgruntled member of the community called the county commissioner and reported that some of the farm's buildings were not up to code and did not have permits. Roger goes on about the inspections that have taken place and the cost of hiring lawyers, eventually arriving at how things currently stand. As he talks his eyes become glossy and he pauses several times when the frustration in his voice gets to be too much. He feels that some of the building codes are developed with the intent to conform the houses of Pavilion township. "Buildings made out of recycled wood with structures jutting out at various angles are not as attractive to the eye as these huge prestigious homes on many acres are," he points out. Roger Ulrich is tired of fighting a point that has fallen on deaf ears. He's exhausted physically, emotionally and financially.

The residents of Lake Village agreed to vacate the "unsafe" buildings, which include in the structures, the cabins out in the woods, the campers, the teepees, the tents, everywhere except the one house where some members were lucky enough to move into. Others were left with no choice but to leave their home, Lake Village, at least for the time being.

On the way back to his house, Roger stops to visit with the animals. First he rubs noses with the pigs, then he feeds the horses some hay, stops in to show the goats, kisses the turkey and pets the cat. It is hard to imagine that this is the same person whose doctoral dissertation was based on shocking monkeys and rats. He is much quieter now than he was when the tour began. It is apparent that talking about the conditions that have befallen

Lake Village is hard for him to do. After walking in silence for almost a full minute, Roger softly chuckles. "I guess I should pay more attention to what I preach in class, to 'expect the unexpected' is much easier to teach than it is to learn."

Lake Village Homestead
A Letter to *The Michigan Land Trust Newsletter*

Reprinted from The Michigan Land Trust Newsletter, *December 27, 2010*

I received the fall 2010 MLT Newsletter: Cultivating Resilient Communities, and I thought it might be of interest to you to share some thoughts regarding those words and MLT's efforts to work with the Transition Van Buren-Allegan group to educate people about the key issues of peak oil, climate change, economic decline and the goal of resilience.

Certainly the issues that you have included as concerns of the Transition Movement are issues that we folks at the Lake Village Homestead farm co-op have been considering since our inception and in fact were the issues I lived with as a youth growing up in the "Anabaptist Amish Mennonite" culture that existed in the area surrounding the towns and cities of central Illinois. Every last one of my relatives and cousins became bankers, lawyers, ministers, politicians and teachers. C. Henry Smith, an older cousin, received a Ph.D. from U of Chicago, and cousin Tim was President at Heston and among other things taught at Blufton. Cousins Willard and Milton, each with a Ph.D., were educators—Willard at Goshen, Milton at Lake Forest. In my Dad's case, he became an owner and manager of a John Deere implement company store. Beginning in about 1924, Dad started working off the farm from which Grandpa C.M. Ulrich with his eleven children made their livelihood. He worked with his team of mules helping build U.S. 24, which went through Eureka. He also hauled peas, sweet corn and pumpkins and other crops from farms to the Eureka Canning factory, and soon after turning 21 years old started working full time at the Eureka grain elevator. This work he did until about 1927, when he married Della Smith and went into business with her father, granddad C.H. Smith, another farmer, selling and trading John Deere tractors and farm implements for money and horses, which were being no longer respected as the "bringers" of energy for tilling and harvesting at the same level as the new GAS GUZZLING Technological Wonders called tractors and trucks. The horses Dad took in on trade were then resold to be made into meat for the French and Belgians, who were less fussy when it came to eating horses than were Amish Mennonites.

Once again, speaking of "Transition Movements," those days defined a Transition Movement beyond words!! A "get-bigger-or-get-out" attitude (before Earl Butts) was fostered by banks, universities, attorneys, politicians, you name it. All of it was a part of what led to an economic crash that by 1936 left my Dad broke and out of the tractor business. The farmers could no longer pay Dad the installments owed for the tractors and other farm stuff because somehow the wheat, oats, corn, soybeans, pumpkins, peas and hay didn't bring in enough money—even when you fed foodstuffs through hogs, sheep, chickens, cattle, etc.

So my other cousins and I went to school; we forgot for the most part how to farm but did well with number 2 pencils, marking True or False, A,B,C,D or all of the above on high school and college tests. We graduated, got cars and "better" jobs; and some of us, like Maynard Kaufman and I, received Ph.D.'s and talked and wrote our way to the innards of academia and back out again onto the Earth, where what we learned from our original family farms began to make more and more sense to us. When I looked around in horror to find that far too many folks had been pushed and pulled into lifestyles through a brainwashing even in our education system (that left Honors College kids at WMU thinking that it hurt the mother hen when the baby chickens nursed with those hard little beaks and that food was "no big deal," something that you get at Meijer's, or restaurants like Zazio's or Big Macs at McDonald's). It was then that I figured we needed another transition.

Thus, starting around 1966, I began to work toward the goal of getting some land to found a community like B.F. Skinner's Walden Two. That all happened in about 1971, and I soon found that I wasn't getting much help keeping alive on the land from Harvard types. Skinner told me he didn't know anything all that much about farming. He had them milking

Herefords in an early, pre-published version of *Walden Two,* until some farmer he knew explained the difference between beef breeds and milk breeds. Also any fool farm kid who has ever been around sheep in any serious fashion would sense that a story where one of the wise elders of the *Walden Two* community explained to visitors that a string was all that was necessary to confine sheep since their grandparents had been raised within an electric wire fence...was bullshit!

I can tell you the early years at Lake Village were rough, especially when it came to raising food, planting gardens, canning, handling goats, sheep, chickens, cattle, pigs (my God, pigs!); it was at this juncture that I began to see that I had to forget about relying so much on the written and spoken word but rather (for heaven's sake) find some people who knew something about the Earth not because of what they said or had written or read, but rather what they could do.

To make a long story short, we found ourselves heading down to northern Indiana and getting in touch with distant Amish cousins who have the best record in the nation for the lowest rate of high school dropouts, i.e., zero. The trick, of course, was that they did not allow their kids to go the H.S. in the first place and thus take the first steps towards the world of Publish or Perish...make a lot of money and buy all the gas and oil you need. I want to emphasize that I'm not against education, I am for it big time. Just don't make it synonymous solely with earning a piece of paper called a School Diploma or a College Degree.

In short, we at Lake Village have been trying to reverse the transition which took my Dad and millions like him off the farm, to where they were all convinced to drive cars and not a horse and buggy, and to plow and harvest with power coming from the oil that we kill for.

Finally, I see that MLT has organized talks, workshops and has encouraged educational and hands-on programs in communities. Most important for me is the question: do you by chance know of anyone who would like to live and work on a working farm like ours, who isn't doing so at the moment and would at least like to try it for a day or so? If you do, have them contact us; we could use some hands in the dirt. In return we have land, housing, food and some cash that we could share with them that hopefully would make their adventure gratifying in more ways than one.

I think it would be a good thing if your group and our Lake Village family were to explore further ways in which we might join forces as we move down the path to still another transition, Armageddon or Heaven, call it what you will.

IN THE MEDIA

Roger Ulrich and Lake Village Community
"The Experiment of Life"
By Deborah Altus

Reprinted from Communities, the Journal of Cooperative Living, *No. 98, Spring, 1998, pp. 52-54*

The land where Roger Ulrich lives in rural Michigan is the perfect community setting: lovely woods, a serene lake, green pastures, attractively-weathered farm buildings, contented farm animals. All that is missing—these days, anyway—is the usual hustle and bustle of people.

But for over two decades this land was the site of an unusual community experiment that touched the lives of scores of people. Now, because a disgruntled former resident forced the local government to red-tag code-violated buildings and evict many of the residents, the 115 acres that housed 30 some people are home to only eight. Ulrich and his wife, Carole—two of the founding members—and six other residents are left to pick up the pieces of the intentional community where they invested 25 years of love, elbow grease, life savings and dreams.

While Ulrich clearly feels frustrated and saddened by the events that have transpired over the past couple of years, he sees no reason to leave the land he has grown to love. He agrees with the advice of Rolling Thunder, his spiritual mentor, who warns that people don't escape their problems by moving away. And thankfully, Ulrich reports, about 25 former Lake Village residents live nearby and continue to offer support to the community.

A framed list hangs in the huge barn converted to living space. The roster includes nearly 300 names of past Lake Village residents—many of them former students of Ulrich at nearby Western Michigan University. For a brief moment, as he reels off dozens of names of former members in the cavernous dining room that could easily seat 50 people, Ulrich looks frail and stooped, the weight of the past year's events on his shoulders. But as he straightens and starts talking about Lake Village, his voice takes on great conviction and fills the room.

Vigorous, trim and talkative, 64-year-old Ulrich spends his days with farm chores, writing, reading, playing softball and teaching and studying Native American and Amish practices. He refers to himself, only partly in jest, as an "Amish Indian." The Amish, he says, and his Amish-Mennonite grandparents in particular, served as his practical mentors, while Native Americans have been his spiritual guides.

"Behaviorist Amish Indian" might be more accurate, though Ulrich would probably wince. At one time, he was a devoted Behavioral psychologist and a colleague of B.F. Skinner and other renowned members of the *Beyond Freedom and Dignity* crowd. Even now, after he has long followed a path widely divergent from his Behaviorist roots, Ulrich peppers his conversations with quotes from B.F. Skinner, along with stories from Rolling Thunder and sayings of his Amish grandmother. One of his prized possessions is a hilarious video clip of him and Skinner discussing everyday applications of Behaviorism as guests on a 1960s Dick Cavett show. It is hard to imagine that the serious young scientist on the TV screen applied his experimental methods, quite colorfully, to his own life.

After finishing his dissertation research on shock-induced aggression in animals (a great embarrassment to the current animal-rights activist), Ulrich pursued a typical academic career—first at Illinois Wesleyan and then Western Michigan University, where he moved in 1965, at age 34, to head the Psychology Department. According to Ulrich, he and his colleagues made WMU into "sort of the citadel of Skinnerian psychology." Oddly, it was Ulrich's attraction to Skinner's book, *Walden Two*, that first got him moving away from Behaviorism.

In the second half of the '60s, Ulrich participated in a series of national conferences focusing on starting a Walden Two community like the one described in Skinner's book. Efforts fizzled to do anything as a group, however Ulrich and fellow participant Kat Kinkade went back to their respective home bases. Both were bright, energetic, colorful self-starters who excelled at doing their own thing. Neither was content, unlike most conference attendees, to just talk about *Walden Two*; they wanted to live it. Kinkade went on to co-found Twin Oaks community in Virginia (and later, East Wind and Acorn).

Meanwhile, back in Michigan, Ulrich and some of his WMU students, attracted both to *Walden Two* and the emerging '60s counterculture, formed a co-op house in Kalamazoo. From there, they put together a non-profit organization and planned a rural commune. In 1971 they bought land next to Long Lake in rural Kalamazoo and formed Lake Village Community.

Lake Village, according to Ulrich, was inspired by *Walden Two* only in the sense that members wanted to experiment with their lives as Skinner, through his character "Frazier," exhorted people to do. Ulrich took Skinner to heart, and in the spirit of the '60s, experimented with many aspects of his life, including open marriage. How did that go? Well, he points out, he and Carole are still married.

Ulrich's students and fellow community members started taking drugs, and he became intrigued. At Harvard he'd attended a presentation extolling the benefits of LSD but was discouraged to learn that the researcher hadn't even tried the drug (the researcher had sophomores take it, then wrote about it). This approach, although standard in academia at the time, was becoming increasingly repugnant to Ulrich—it seemed dishonest. He wanted to learn from direct experience. And his willingness to experiment on hapless sophomores—especially when he lived in community with them—was waning. Skinner's words shouted to him from the pages of *Walden Two:* "You have got to experiment with your own life, not just sit back in an ivory tower somewhere…."

So Ulrich took drugs. Lots of them. He started with grass and moved to psychedelics. After a particularly frightening trip dropping acid and shooting cocaine, he forswore mind-altering chemicals.

Meanwhile, Lake Village had become a happening place. Students and other members of the hip underground moved into every possible nook and cranny. Farm outbuildings were renovated to house the burgeoning population; pastures and woods sprouted tipis and domes; wondrous, artistic, cozy dwellings appeared in sheds once meant for chickens and cows. Peter Rabbit of Drop City fame remarked that it looked "like a group of people trying to come down from a middle-class trip."

Nobody paid much attention to building codes. In fact, little thought was given to law and order—external or internal. In the spirit of the times, freedom was greatly valued, and, for the most part, people were able to do their own thing without getting in anyone else's way. In the early days Lake Village did have meetings, committees and planners, following a rough *Walden Two* outline. But people quickly decided that they were doing too much talking and not enough doing and replaced weekly meetings with "Work Sundays." Ulrich points out that the community had its share of relationship hassles and spats over work-sharing. But, he says, that happens any time you get people together; it wasn't peculiar to their community.

Early on, members tried to use some Behavioral methods with their children. Ulrich laughs and shakes his head when he recounts their experiences with a *Walden Two*-style system through which the kids earned points, exchangeable for privileges, by doing chores. Their earned points, recorded on wall charts where they could see them mount up, became "positive reinforcement." The problem with this system, he says, was that the parents paid too much attention to charts and not enough attention to their children's actual behavior. The point system was abandoned after they discovered that the kids were earning points for cleaning even though they were sweeping cat litter under the carpets. Ulrich, the supposed Behaviorist expert, was humbled. His book series, *Control of Human Behavior*, cited study after study that demonstrated effective behavioral control—"where everything turned out right, just like in *Walden Two*." But, as Ulrich said in a 1978 *Communities* magazine interview, "Now here I am living in this commune, and I can't get my kids to put away their socks."

Ulrich freely admits that he has played a strong leadership role at Lake Village over the years. And he recognizes that, at times, he has pushed his authority a bit too far. Some members clearly resented Ulrich's influence. As one frustrated member said in a 1992 *Kalamazoo Gazette* article, "We want to make a group decision but whatever he [Ulrich] says goes." Yet others argue that Ulrich was a reluctant authority who fell into that role because he'd been at Lake Village the longest. According to another member, "Roger has said he wishes he could give everybody a '20-year pill' so they could get caught up. It's really hard for him to be the leader, but people there have to focus on how they can get involved. He has always wanted them to do that."

When asked if he had a vision for the community, Ulrich responds that it's best summarized by the saying, "Do the right thing." During the 1970s Ulrich tried to do the right thing by offering Lake Village as a sanctuary for refugees from other countries. Over the years Lake Village has been home for a number of refugees from Cuba and Central America. The community went out of its way to help these refugees even though it assumed a certain amount of risk by sheltering people without official immigrant status.

But risk is something that Ulrich has never shied away from, then or now. Living in flagrant disobedience of building and zoning codes was risky, and now, after more than two decades of doing so, Lake Village is paying the price. While clearly frustrated by the siege on his community, Ulrich is far from defeated. He feels no desire to jump ship, especially as he remembers Rolling Thunder's words: "You have to live the truth and be a part of it, and you might get to know it. I say you *might*—and it's a slow and gradual process and it doesn't come easy."

Indeed, life is not easy now for Ulrich. He worries a great deal about what he sees as the exploitation of the Earth's resources, both in the Kalamazoo area and globally. "With ever more oppressive county codes, people move out of cities onto land that is no longer affordable. This especially hurts traditional people who have historically lived close to the Earth." And, Ulrich continues, those who have typically respected the natural world are forced by laws that exploit natural resources into a life increasingly detached from the Earth.

But despite his considerable concerns, the gleam in Ulrich's eye suggests that Lake Village may rise again. His feelings about the future are perhaps best summed up by his last words in a 1970s magazine interview—"May I express the conviction that the experiment is not over?"

Field Trip Helps Cooper Students Learn about Agriculture

By Sharon Sturdevant

Reprinted from The Kalamazoo Gazette, *May 27, 2008*

PLAINWELL—Science lessons for some first-grade Cooper Elementary School students looked a little different recently when the kids headed on to the Lake Village Homestead Farm. Gone were the books about animal life cycles and the worksheets about how plants grow.

Instead, the 47 kids and 17 parents were busy collecting eggs, planting potatoes, milking goats, seeing a 2,000-pound bull and riding horses.

"The students were so engaged and enjoying themselves from the moment we arrived," said Cooper teacher Kim Hannig. "They were able to be kids while learning and experiencing things."

This was the first time that Hannig and fellow first-grade teacher Sara Smith took their classes on a field trip to the cooperative farm tucked at 7943 South 25th St., along Long Lake in Pavilion Township. But the trip, suggested by farm resident Tony Kaufman, whose daughter Ella is in Hannig's classroom, provided a strong connection between science topics the students studied this year and the real world.

"There are still a few farm families in our school," Hannig said. "But many of our kids live in new neighborhoods being developed between Kalamazoo and Plainwell. The students just don't have as many opportunities to explore the natural environment."

Agrarian experiences are commonplace for the 15 or so individuals who live on the Lake Village Homestead Farm and other people who live in a dozen homestead properties surrounding the farm. The group sells associate farm memberships that entitle area residents to participate in activities like managing livestock, gardening, hunting and fishing on the farm and to purchase foods grown or made on the farm at reduced prices.

But those involved with operating the farm believe strongly that children who experience agrarian culture and explore the natural environment gain valuable skills and knowledge.

"Pediatricians say that kids are healthiest and happiest with more free, unstructured outdoor play time," Kaufman said. "Plus, what kids see is what they will become. Kids need to see local, adult people working hard to encourage the value of small family farms producing food in healthy ways to keep a community healthy. The kids need to get outside and see the elements that actually keep them alive."

Kaufman, who has lived for about nine years in one of 12 homestead properties surrounding the farm, told Hannig about how various groups visit the farm to participate in interactive farm workshops and encouraged her to bring the students for a visit.

"It's not that we won't bring animals to the schools," Kaufman said. "But kids need more time in nature and natural settings where food is raised. That's where some of the real learning occurs."

A donation by the school's parent club organization enabled Hannig and Smith to take the first-grade students to the farm on May 15. The students were divided into three groups that rotated through activities in the barn, garden and horseback area, Hannig said.

"The groups were so busy with activities that I didn't even see all the things they got to do," Hannig said.

At lunchtime, the group climbed aboard two tractor-drawn wagons with sack lunches and headed through a natural area into a pasture where 40 or so cows grazed with their calves.

"I thought he (Kaufman) was kidding when he said we were going to have lunch with the cows," Hannig said. "I was really surprised when he stopped the tractor in the middle of the field. But the kids loved doing that and then drinking some of the fresh goat milk."

The students also got a chance to run through the fields and hike along the wood trails, Kaufman said.

"A few kids got to step in mud and a few others got to step in cow pies," he said.

Some of those details showed up in stories the students wrote after the field trip, Hannig said.

"The kids were so engaged in that writing process," Hannig said. "They started writing without hesitation and didn't complain once that they had nothing to write about. Their excitement about the learning experience definitely showed."

Is It Utopia?
By Maureen Aitken

Reprinted from The Kalamazoo Gazette, *September 9, 1992*

Delilah screeches and writhes but Manuel Garcia carries the pig from the half-finished coop and places her on weeds still stiff from the summer's cool morning.

No pigs in the chicken coop, Manuel lectures, but Delilah just snorts at the weird little shack being made from the roof of a dead pickup and an assemblage of used and new wood. A few tired commune members slouch on a nearby hunk of metal and they wax on about the importance of recycling and Mother Nature until the pig comes back.

"Pigs are smart, man," someone says, watching Delilah inch closer to her pig friend, Samson. "Smarter than dogs."

The animals, like the people on this nearly 325 acres on Long Lake in Pavilion Township, seem to creep up everywhere between the recycled hunks of metal, pieces of cars, and discarded Western Michigan University chairs that surrealistically appear to sprout like wood flowers around the horse fencing, behind the barn, out in still fields.

For 21 years, the people at 7943 S. 25 St., have wanted to live in the palm of nature and outside the typical house, kids and picket fence way of life. But continual ego clashes, lack of money and practical concerns are causing some newer members to say the commune needs serious changes.

"Communes can work, without question," said one frustrated member. "But this isn't really a commune, it's just apartments in the woods. They use the commune ideal so they can get a lot of free work out of you."

Each of the 300 members who have called Lake Village their home have wanted a chance to bridge ideals with reality. The commune was based on an idealized, fictional community created by B.F. Skinner in his novel *Walden Two*.

Lake Village's founding members were in reality Western Michigan University Professor Roger Ulrich, his wife Carole, and, in name, his father. Although a group of intellectuals joined Ulrich in the effort, he and Carole remain the sole founding members living on the commune.

Everyone on the commune knows Ulrich's story—raised in a Mennonite family, he received a doctorate in behavioral psychology which entailed shocking monkeys in his study of pain and aggression.

At Lake Village, Ulrich has tried to mold a hybrid of Mennonite, Native American and '60s ideals.

Today, Skinner is interpreted along Long Lake by about 20 people living in converted barns, houses and other buildings on 115 acres of the larger piece of property. All members must participate in chores, and some are paid up to the maximum $4 an hour for extra work. Most have outside jobs and practice no spiritual beliefs other than respect for the land and militant recycling. Residents pay rent, or "shares" plus parking and other costs, but basically, their money is their own.

Another 20 live on their own property along the commune perimeters. The commune also has a connection to the Learning Village at 202 Fairfax.

Living Skinner's ideal has not always been effective or easy, Ulrich admits, and the commune has had its share of drugs, suicide attempts and apathy. Time, too, has worn down some of the idealism and an alternative lifestyle still entrenched in the 1960s.

But the commune has brought great satisfaction to members like James Boshoven, 30. Boshoven was feeding Buddy, a young bull as he talked about the stereotypical view of communes as drug farms and sex parties. That is just the fear expressed by people who can't see beyond their brick homes and primped lawns, he said.

After a Peace Corps assignment in West Africa, Boshoven came to the commune because he had already been mugged in Kalamazoo. He had seen how people talked little and threw away a lot.

"To me the most important thing offered is personally developing skills in gardening, animal raising, husbandry and homesteading," Boshoven said. "It's engaging in a community that works together to solve issues."

Boshoven and about 20 other residents share daily chores that require hard work and often early hours. Between feeding animals, milking, cleaning, repairs, gardening, home construction and everyone's outside jobs, there's little time for serious debauchery, according to one resident.

"There are more drugs and sex at the university," said Glen "Phrede" White.

Residents say there is a peaceful calm in watching sunsets along the vast stretches of land, in seeing the morning light bead along Long Lake, and in seeing the new life emerge each spring. There also something vital, they say, in co-existing with people who have vastly different personalities.

"If there is one common goal, it is respect for nature and each other," said Roberta McInnis.

McInnis, one of the younger members at the commune, said she joined when she found she was not getting what she wanted from the traditional

Kids and kid pigs meet at Lake Village. For many children, for whom their closest encounter with pork is at breakfast in the form of a slab of bacon, this experiment is really very, very new and exciting.

Kalamazoo lifestyle. She wanted to be able to taste her own tomatoes, bike along country roads and feel safe.

Unlike McInnis, sometimes younger members can be overly idealistic, but they soon learn all is not breezy and sweet down on the farm, Ulrich said.

"The romance brings them here," Ulrich said. "Then the deflating of that balloon starts to occur and those that stay are the cream of the crop."

But the younger members are part of the national revitalization of communes like Lake Village, according to one national expert.

While communes fell out of favor in the late '70s and '80s for being too chaotic, philistine and impractical, today's younger people wanting to merge environmentalism and community are showing up at the doors of communes like Lake Village. These communities are experiencing a resurgence like never before, according to Laird Schaub, who studies intentional communities across America and is based at one such community in Sandhill Farm, Rutledge, Mo.

"With workplace democracy and participatory government, I think the interest is growing dramatically," Schaub said. "In my 15 years of being involved in intercommunity contact, there's never been a greater interest in joining or starting communities than now."

But the young are also raising questions at the commune that older members have not thoroughly considered, some members say. The biggest questions are about the practicality of so much work for so little money.

Many spend their mornings milking goats, tending fields, cleaning stables and building for little or no money. Commune members are supposed to remember the common good when early morning fields look like frozen tundra and the pigs have to be fed.

Benito, who had to flee Cuba many years ago now, has only limited control of his shaking hands since a stroke. He and Manuel Garcia fled Cuba during political turmoil in the 1970s. Manuel has since had to leave the commune to find a cheaper place to live. An expert carpenter, Manuel builds most of the projects at the commune and makes the commune maximum—$5 an hour, although Ulrich says Garcia can make more on bidded projects.

Garcia looks for other work, which is hard to find for a man who speaks broken English.

"Lots of work here," he says, from the top of his creaking wooden scaffold, where he has been pounding away and talking to himself in a bellowing tone. "But no money to pay."

White, like some others, has growing concerns about his situation. He pays $250 for a room and a common bathroom. In the winter, he paid to take an independent study as part of his WMU psychology degree. He worked the agreed 260 hours gratis, but then Ulrich said White needed to work more.

Although Ulrich does not have assistantship funding through WMU, he recently assigned one doctoral student to the commune.

The commune's 115 acres is owned by the Lake Village Behavioral Development Corp., in care of Ulrich, one of seven directors on its board. Some older commune members own small chunks of property, but Ulrich's name is on some of the outlying land, according to Pavilion Township tax records. The entire 300-some acres has a $7,600 property tax bill for 1991.

Ulrich, who says base shares for rooms can run as low as $90, said rent costs partly pay for privileges of the recreation room, the land and other facilities. Ulrich also says he paid $3,000 out of his own pocket for taxes last year.

Many of the younger members with financial concerns say it's time the commune became financially productive, as other communes around the country are.

Members are trying to find ways to market and sell more crops and products derived from cows, goats and other animals, according to commune member Craig Fitch. But it's difficult to change old attitudes that don't embrace the concept of profits.

"A lot of changes are possible," Fitch said. "We have four times as many gardens as last year… This is an old place, like a big ship that has to be turned around. But we've got a good group of people, and they are more involved with the land and the community."

But others, like White, are questioning authority that has been established by Ulrich over the past 25 years.

White is still working and living at the commune but feels nagging doubts about the supposed "communal" atmosphere.

"He does make it seem that way," White said of Ulrich. "But there are verbal games. We want to make a group decision, but whatever he says goes."

Ulrich, for example, summarily canceled a musical benefit at the commune that eventually did take place a few weeks ago. He has also pushed people out of the commune by using mind games, White said.

When talking about his own life and others, Ulrich can take on a preaching manner.

His conversation is peppered with catchy phrases such as "think globally, act locally." And Ulrich sees himself as someone who easily relates to others.

People, Ulrich said, are addicted to caffeine, television, money, general consumption and various other drugs. But if he can give up a series of vices, so can anyone.

"We're all junkies," Ulrich said. "We're all addicted to consumerism…. I have all of the problems anybody else has. There probably isn't any substance I haven't used. I've shot the s--- in my veins. I know what I've done to my body."

Ulrich may come across as the heavy, but he really is better at seeing the broader perspective. He then becomes a reluctant authority, said McInnis.

"They've been there 20 years or more," McInnis said of Roger and his wife, Carole Ulrich. "Roger has said he wishes he could give everybody a 20-year pill so they could get caught up. It's hard for him to be that figure, but people there have to focus on how they can get involved. He has always said they could do that."

Because Ulrich has been living at the commune the longest, he agrees to some authority, although with hesitation.

My free-spirited daughter Traci riding in winter.

"Over time it will probably be my rule, or my lot in life, to say 'you really ought to clean that up,'" Ulrich said. "It's more likely I have to get a little stern, but if you hear the story from the person you got stern with, it's like hearing from the child that got pissed at daddy."

Those who come to the commune sometimes suffered difficult childhoods or complications that cause them to question typical society or authority, Ulrich said. And naturally they would question his authority as well.

"I understand it is very good for someone to lash out at me as a sign of growth that needs to take place," Ulrich said.

Maybe one of the most nagging questions raised by younger and older members at the commune, however, is ownership. Even if residents work 20 years on the commune, they own nothing, because in essence, all humans are renting the earth, according to Ulrich.

But for McInnis, the stress of multiple demands at the commune and helping to build a house that will never be hers raised a lot of questions. McInnis and her boyfriend, David Zimmerman, moved out to a house in South Haven this summer.

"It's hard to describe how I was feeling then but I'm much more at peace," McInnis said from South Haven.

Others have chosen to buy land along the commune perimeters. Robert Brown, who now lives with his family at 7314 25th, thought when he was young that he could live at the commune indefinitely. But age and family made him buy property adjoining the commune.

"There's more privacy and more space with a place of our own," Brown said. "We wanted to basically have something we could call our own."

Although he and McInnis still participate, they are able to feel their futures are more secure.

Ulrich, recognizing the practical limits of communal living, has sold two pieces of property along 25th Street, and others who are thinking about long term residence at the commune are starting to wonder if ownership is better than renting.

The idea of ownership as opposed to contributions is taking hold on many communes in America, Schaub said.

"In these communities it's certainly not unusual to change its relationship to property ownership," Schaub said. "It's more common to get individualistic rather than more communal."

While Ulrich does not like the idea of selling off the 115 acres of Lake Village, he can see people continuing to buy around the perimeters. Ulrich said he even considered buying his house at Lake Village, but thought it went against the idea of community spirit.

But the concept raises one of most difficult questions facing the commune. Without selling, people face an uncertain future, but doesn't selling pieces of the commune go against the unified purpose of communal living?

"It's a good question," McInnis said, taking another drink off her beer and watching the sun burn crimson against the fences, the crops and all the wood chairs.

Associate members Marcela and Samuel talking to Paco during a three-day visit to Lake Village.

Lake Village Homestead Lives On
By Donna Allgaier-Lamberti

Reprinted from **The Kalamazoo Gazette,** *August 21, 2000*

When Roger Ulrich was a young psychology research professor and director of the Behavioral Research and Development Center at Western Michigan University, he conducted experiments on animals to try to discover a way to control human aggression. He later realized that he was looking for answers in the wrong places.

As he later wrote, "Experiments do not cause things to happen. University classrooms and laboratories do not represent a natural setting. There is no experiment other than the real situation."

Experimenting has become a way of life for Ulrich and participants of Lake Village Homestead, an intentional and cooperative community. Convinced of its merits, he and his wife Carole brought their son and two daughters to the farm to grow their own food, raise their own animals and participate in the community-based living experiment.

Ulrich founded the Lake Village Homestead and opened the Learning Village, a publicly utilized preschool day care center, under the umbrella of the nonprofit Behavior Development Corp. in the early 1970s. Inspired by the behavioral model outlined in the 1961 utopian novel *Walden Two*, written by B.F. Skinner, Ulrich bought approximately 120 acres of land from Bill Bethel, a former farmer.

Located at 7943 S. 25th in Pavilion Township, Lake Village Homestead is a diverse collection of individuals and families. Co-housing and single-family dwellings form a community of people working together to create a sustainable, rural ecovillage.

Associates live and practice an environmentally conscious lifestyle that involves an ecological, environmental and social way of living while staying conscious of the needs of its members and the land.

The community is overseen by a six-member board. The board sets association policies after seeking input from its residents, and is dedicated to preserving open space as well as natural wetland habitats.

"We are a loosely knit confederation with differing degrees of cooperation," Ulrich says.

Once rooted in the back-to-the-earth communal practices of the 1970s, the community has matured into an environmentally conscious lifestyle of people practicing what sociologists call "right livelihood"—living and working in an authentic way that is based on one's belief system.

Simple life

Ulrich grew up with an agricultural background, spending summers on his grandparents' Mennonite farm. It was a way of life that emphasized living well using few resources, along with the belief that small is beautiful.

"Some of my peers at the university had a hard time understanding why I would choose to live here in such simple surroundings," Ulrich says. "But I think of this farm as my paradise."

Ulrich often finds himself acting as guide, protector, adviser and friend to those who live in the community. It was his vision and hard work that provided the roots from which the community has grown.

"My role is like that of an elder in an old-order Amish community," Ulrich says. "Although spiritually I connect more to the indigenous Native American culture."

Sense of harmony

On the working farm at Lake Village Homestead, residents raise livestock, grains and vegetables for food and income. They also raise ducks, chickens, pigs, turkeys, goats and cattle for food, using their byproducts to fertilize the community's large organic vegetable garden.

All animals are free range—free to roam in their pastures—and all gardening is done without the use of synthetic chemical fertilizers, pesticides or herbicides. Eggs and meat produced on the farm and not consumed by residents are sold as a source of income.

Lake Village Homestead is made up of 320 acres of forest, meadows, wetlands and farmlands. Some of those acres form a licensed pine tree farm; 120 acres are planted in hay. The hay is used by Lake Village Homestead residents for gardening and animal needs.

Within Lake Village Homestead is a unit called "The Co-op"—a cooperative of various homesteads. Approximately 120 acres of lakefront on Long Lake fall under the umbrella of the nonprofit corporation, and are surrounded by land that is owned by the 11 families that reside there, many of whom at one time or another lived in the homestead.

In keeping with the cooperative philosophy, village members and neighbors share equipment and work together on farming activities. For example, when plowing and harvesting need to be done, neighbors who live nearby help with chores in return for food.

Teaching a child to drive Aries in harness is as much fun for me as for the kid... maybe more.

Human connection

Members who live in Homestead are proponents of the simple-living movement. They share housing, vehicles and farm equipment, and they work cooperatively for the common good of all residents. Lake Village focuses on sustainable farming practices, preserving its land and ground water, growing organic foods and exploring what it takes to live a life within a cooperative and sustainable environmental ethic.

"We consider ourselves servants of the land," Ulrich says. "It is a privilege to live here in this beautiful place, and I thank God every day for the opportunity to preserve this land."

Currently, six residents live at the Homestead, in what is described by the township code as a single-family farm housing unit. Residents have several options for lodging within the community. They may choose to live in the shared housing center, where they sleep in a community bunk area, or they can reside in individual rooms within the common house. The lifestyle in this passive solar group house is much like that at the turn of the 20th cen-

Tony Kaufman's daughter, Ella Timmerman Kaufman, grew up and still grows at the Lake Village farm along with goat kids.

tury, when a number of unrelated people lived and worked together on family farms.

The rest of the participating 11 families reside in houses scattered throughout the village. This is an open group and people flow in and out as needs change.

Because there is only a certain amount of land available, and the board is committed to keeping this land from being further developed, units of land are passed from one individual to another as need and availability arise.

"If one landowner leaves, someone else has the option to purchase the land and structure," Ulrich says. "Many, but not all, residents start out at Lake Village first."

Residents are charged $300 a month, plus their participation in farm chores, for what is called a share. Each contributes to the upkeep of the shared community. The pooled income goes to pay off property taxes, assessments, utilities and other farm expenses. Each resident is responsible for his or her own personal income taxes.

Daily life

Residents work both on and off the farm, some in conventional jobs. Others, like Homestead's current manager, Tony Kaufman, work on the farm in a management and educational role.

Those who work outside of the Homestead come and go much as outside residents who live in the city. Kaufman's day resembles that of a ranch manager, whose job changes with the season and who responds to needs as they arise.

Regardless of their daily activities, participants at Lake Village Homestead share a set of core values that include being committed to a way of living that uses less and tries to remain in balance with the Earth. "We work hard to strike a balance between what it takes to live and what it takes to sustain this land," Kaufman says. "We are very concerned with urban sprawl."

Living laboratory

In the farm's early days, it was a laboratory for the Behavior Research and Development Center. Ulrich brought his students out to the farm, where they used their experiences as well as their observations as material for classwork and thesis development. Now that Ulrich is retired, the farm's use as a university classroom is limited.

Classes and workshops are still offered at the homesteading school, with topics such as composting, naturalized landscaping, ground water protection and organic gardening. Consulting services are offered, and memberships are available for gardening space and use of the nature trails.

Outside organizations, like the 4-H and the Explorers Club, work in the garden, feed the animals, enjoy a hayride to the wetlands and participate in workshops, activities and demonstrations.

Becoming political

Being a political activist has become a way of life for Ulrich. A member of the American Civil Liberties Union, Ulrich says that much of residents' political efforts are fighting for causes that deal with Earth-based issues, such as ground water protection and organic practices. Ulrich realized early in his life that his task involved changing minds as well as molding opinions.

"You cannot be on the fringe of society, be an environmentalist and a social activist, and not be political. Today you hear many people saying, 'In order to save our cities you have to save our farms,'" Ulrich says. "I think they have it backwards. I believe that we need to save our farms first."

'This Is a Real Farm'
Village Cooperative Taps into Health Trends
By William R. Wood

Reprinted from The Kalamazoo Gazette, *May 15, 2006*

Twelve children and a few adults entered the dim, cluttered farm kitchen, not knowing they were about to partake of a culinary surprise.

Plates dotted with bits of cheese and small chunks of hot dogs and salami sat on a table. Paper cups filled with milk were lined up on the table beside the plates.

The cheese was made from the milk of cows and goats on the farm, Lake Village Homestead Cooperative of Pavilion Township. The hot dogs and salami were made from cattle slaughtered there. The milk had come from a goat in a nearby barn earlier in the day, had been put through a filter and then cooled.

This was artisan fare, made without chemicals or hormones.

The milk was the big hit. The adults present to look after the children were as pleasantly surprised by the milk as the kids.

"This is good!" said adult Addie Warren, pulling a cup of goat's milk from her lips after taking a sip. "I was afraid to taste it at first, but it really is good," she said to several other adults in the group, who had been focused on serving the children. "Try it, and you'll never want regular milk again."

Visits like the one by this group from Kalamazoo's Learning Village, 215 Lake St., are becoming more common at Lake Village, 7943 S. 25th St., as Lake Village reaches out to the community to let the public know about its services.

For a membership fee of $200, families can come to the farm and spend time with the animals there—horses, cows, goats, chickens, pigs, ducks, turkeys, cats, dogs and peacocks. On prearranged visits, those who are not members can buy food that the farm produces, including steaks, chops, chickens, soup bones, ground beef and sausage as well as eggs, wildflower honey, maple syrup, vegetables and popcorn.

Lake Village, eight miles southeast of downtown Kalamazoo, is a working farm and educational center for all ages that offers the opportunity to experience all aspects of farming, including building construction, livestock management, gardening, hunting and fishing.

"This is a real farm, not no show stuff," said Nate Butler, another adult with the Learning Village group, as he took pictures with a cell phone of his son Cameron and other children who rode ponies, played with newborn chicks and milked a goat.

"This is the furthest thing from a factory farm you'll ever see," said Roger Ulrich, president of the Behavior Development Corp., a nonprofit group that serves as an umbrella organization for Lake Village.

Kids looking at pigs. Pigs looking at kids. Who is studying whom?

Back in the 1970s, a commune that emerged on this 300-acre section of forest, meadow, wetlands and farmland was not a popular subject in Kalamazoo. The people who lived and worked there were called "hippies."

But both the people on the land and those who once had negative views of them have changed with the times.

The 60 people who help out at Lake Village Homestead, some of whom established the People's Food Co-op in downtown Kalamazoo, are now part of "the establishment." Some are parents who worry about the influence pop culture will have on their kids.

About 45 people connected with the farm who live in houses that ring the property include doctors and educators, roofers and concrete workers. Another 15 people live on the farm.

"This is kind of like coming out of the closet for us," said Ulrich, as he walked down a dirt path on the farm, explaining Lake Village history.

Lake Village is just now spreading the word about its services because many people here, like others across the nation, have grown receptive to the things Lake Village people have recognized for years—the benefits and importance of organic foods, milk free of hormones, locally produced foods and the exposure of children to farm life.

Some of the foods at Lake Village are what some would consider expensive. On the day of the class visit, items in a Lake Village freezer included a whole, five-pound, free-range chicken for $22, a pound of ground beef for $6, two pounds of beef soup bones for $6 and a pound of homemade hot dogs for $8.

Yet the hot dogs alone, slender and long, with a thick texture similar to that of salami, impressed those who came to the farm from the Learning Village. Both adults and children enjoyed the flavor of the hot dogs, which were eaten without having been warmed through.

Lake Farm supporters don't mind the higher prices of the farm's foods because the foods reflect their views of how humans should treat animals and land and how people can live healthy lives, Ulrich said.

It may be trendy now to talk about healthy eating and being conscious about the environment, but that doesn't mean people follow through with actions, Ulrich said.

"There's a dysfunctional gap between our thinking, writing and actual behavior," he said.

He's proud that those associated with Lake Village Homestead not only talk the talk they walk the walk.

Honesty, Respect among Lessons in Living
By James R. Mosby Jr.

Reprinted from The Kalamazoo Gazette, *November 22, 1992*

They came back to back.

One at night, the other the next morning.

Two classes, two teachers, two experiences, worlds apart. Or were they?

One night last week, college students, 80 strong, packed into an unorthodox classroom in the basement level of Western Michigan University's West Hall for three hours.

The next morning, high school students, 30 of them, gathered in an unorthodox class in the John F. Kennedy Center near downtown Flint for long beyond the bells.

Twelve hours, 130 miles and a few years apart.

The college students were taking "Toward Experimental Living," psychology 364, taught by Professor Roger Ulrich, he of Amish roots and one of Western's more involved and innovative teachers.

The high schoolers were specially selected from English, journalism and graphic arts classes and made up into a room run by Flint's teacher of the year, John Ribner, the accent of his native New York thick as ever, and likewise involved and innovative.

Their common denominator?

They had to put up with listening to me. Not only that, they had to hear my life's story, not by choice, but by request. I'd rather talk about the *Gazette* or journalism any day.

I'd been invited to address both groups, by Ulrich and by a mutual friend of mine and Ribner's, with only slightly different purposes.

Ulrich wanted me to talk about what I do for a living, and how I got there, from Day One. The life story of someone in a profession. And then throw it open for questions.

Ribner's students, 16 to 20 years of age, were also to hear about this business and what I do, but with the emphasis on the personal commitments it took to get there, especially when I was their age. Plus time for questions.

Ulrich's collegians had leapfrogged several of life's hurdles already. They were in college, first of all. Most people aren't. And they were well along the collegiate path, beyond the beginnings that defeat so many.

The students at the alternative high school had overcome, or were in the process of overcoming, a variety of their own hurdles. Trouble at their home school. Drugs. Teenage pregnancies. Who knows what?

Most of the college students indicated they were working to put themselves through school, or at least help their folks with the finances. Yet others, older adults with children, were coping with returning to school.

The high school students were struggling mightily to crawl out of the cracks, some with parents, some with one parent, some without either and living where they can.

Of course, the collegians write reports for their class.

Meanwhile, the high schoolers reported stories for the new school paper *The Cougar*.

The Western students were to have read B.F. Skinner's *Walden Two*.

The JFK students were studying the bear in one of Owosso author-conservationist James Oliver Curwood's tales.

In both classes, the questions came from bright kids, wearing their curiosity on the sleeves. Thoughtful, penetrating questions, many prepared ahead of time.

"Would you print the name of the woman who alleges Magic Johnson gave her AIDS, if you knew it?"

"How much do you make?"

"Is psychology a good thing to take if you go into journalism?"

"What do you expect to be doing in 10 years?"

You tell me which class asked which questions.

As Ulrich wrote me beforehand, he tries to get his students "to carefully observe their own actions, feelings, hopes, fears...lives as they relate to others and everything else around them."

His course syllabus says, "Listening, observing and paying respect will be emphasized."

He instructs, "Above all be honest."

Across the state, I'm told Ribner, whose award came from his peers, puzzled people because he asked to be reassigned from a junior high to the inner city community school this year.

But he knew what he was doing. In a few months, this intense, caring professional has already become a father figure in lives with few fathers.

He observes actions, feelings, hopes and fears, lives really.

He urges honesty.

He gets respect.

Experimental living, indeed.

CHAPTER 5
Sustainable Living

Roger is passionate about the importance of sustainable living and the importance of preserving the countryside, as is clear in these provocative titles in his writings about the negative effects of urban sprawl on farmland: *Treating Food as a Friend* and *We Can't Eat Money*. In a 1972 speech reprinted here, Roger exhorts: "At every level of our individual daily personal actions, we must begin to rise to the challenge of living our lives using fewer resources."

◀ *A free-ranging mother pig without her piglets.*

Treating Food as a Friend

Reprinted from the PsyETA Bulletin, *Spring 1988*

In response to the PsyETA call for comments, I would like to make some prompted by the fine article, *We Don't Eat Our Friends: A Report of Research in Progress on Vegetarians* by Amato and Partridge (Spring 1987).

The Animal Rights Movement is justifiably gaining strength around the world and needs to focus its energy so as to avoid needless hostility toward individuals who share concern for other life forms, albeit via different methods of expression other than vegetarianism. For example: in a recent letter to the editor of an animal rights journal one reader wrote, "I am convinced that true enlightenment regarding animal rights cannot be achieved while one is still carnivorous. The eating of meat is both a sign of halted moral evolution and a contribution to the stagnation of personal evolution." The editors replied, "We agree that vegetarianism is essential to the practice of a sane animal rights ethic." Pretty strong stuff aimed at folks who love animals and still eat meat.

Like many other animals. I consume foods that were once alive. I even eat squirrels, rabbits and deer that cars have hit. Deteriorated road kills, of course, go directly to the Lake Village Community's pigs, dogs and chickens, who seem less fussy about what is in their diet than do most humans. Some winters back, Phyllis, an old cow, slipped on the ice. When we determined she would never rise again, I shot her, and nature's recycling continued. My favorite horse, obtained years back from the Upjohn Company, after they were through experimenting on him, killed himself on a sharp board in our corral. Black was 25 years old and had been living on the commune for 13 years. After he died I lay my head on his and prayed that humans might better understand we are not superior to other life forms. Later, I skinned and put some of Black into a freezer and from time to time took sustenance from his remains. The rest of him lay beyond the pines where the dogs and wildlife ate from him all winter.

We drink milk from our goats and use it also to make cheese. We eat the eggs from our fowl and if one dies we feed it to the pigs. Sometimes we eat bacon with the eggs. Countless creatures are right now making a meal of me and I, like you, will one day die and the process will continue. To remember that all my food represents life, of which I am a part, and to work at not taking more than my reasonable share is, for me, the most important thing.

Life is a question of balance of which we are all a part. We are a maze that we call human and almost all the substance of this maze was once other living bodies of plants and animals obtained by something first causing death. As Alan Watts notes in *Murder in the Kitchen,* "All of us are other life forms rearranged." After seeing the *The Animals Film* at the Chicago Film Festival, I was invited to a vegetarian meal. There on a long table, high above the city, lay once living things grown from the labor of humans and animals; carried, carted, trucked, railroaded, shipped and flown, from all over the globe to Chicago, so that a small portion of the 5% of the world's population that yearly devours 45% or more of all the Earth's resources could eat in vegetarian style.

Humans, regardless of what they consume to keep themselves alive, are the most unabashed wasters that have ever lived. We are fouling the surface of the planet as we burn the fuels to grow and bring us special foods. We are destroying animals, birds, fish, insects, fresh water, air and earth. We convert almost everything we touch into cities, suburbs, sewage, smog, roads, rust and ever enlarging fields upon which big tractors inefficiently roam to grow more things to eat. Meanwhile at schools, churches, scientific and other conventions, we insanely preach of our enlightened sanity and ascendency over other life forms.

Certainly many people greedily consume far too much animal flesh. Exclusive killing and eating of plants, sometimes promoted by vegetarians, may simply be another form of insanity which tends to mute the screams that accompany the vegetables' trip to richly laden human tables. Most certainly it alone does not constitute an answer to the very real suffering of laboratory and other animals. There is no way to avoid the fact that life feeds on death. Most importantly, all life forms deserve respect and all forms of energy consumption fall under the Second Law of Thermodynamics, the law of entropy, so well laid out by Jeremy Rifkin.

Therefore, let us look to our total lifestyle as we repair our human faults. Would it not be better to eat more from that which has been husbanded, mothered, cherished and sacrificed within the context of love, be it flesh or vegetable, than to feed exclusively from the stuff purchased in many modern stores, killed, canned, boxed and sacked within the embalming materials listed in chemical jargon on their paper labels?

Sam, Della and Ella with goat kid friends.

We are what we take in and put out, and nothing seems to change as fast in nature as can the computer type these words. The problems of humans are those faced by all life. Greater temperance in our eating habits is a good step for us to take and also to realize that we are hunters engaged in stalking and stocking our food. Some people hunt with guns in places where nature is more similar to times past. Others hunt through the ads on television and in the newspapers and listen to the sounds of voices over the radio—which tell them where to drive to bag the best bargain and how one car with a certain type of tire, oil and gas is better to hunt from if one wishes to look sexy and cool while in fast pursuit of health and happiness. Human hunters should be careful when pointing a finger at other human hunters…after all, each and every one of us is part of the biggest problem on Mother Earth… The Human Race.

Human and Plant Ecology: Living Well with Less

Reprinted from People-Plant Relationships: Setting Research Priorities *(ed. Joel Flagler and Raymond P. Poincelot), The Haworth Press, Inc., 1994, pp. 193-201*

This paper is a modified version of a presentation given April 24, 1992, at the Meadowlands Sheraton in East Rutherford, NJ, as part of the People-Plant Symposium, "Setting Research Priorities," sponsored by Rutgers University, the People-Plant Council, the American Society for Horticultural Science, the American Association of Botanical Gardens and Arboreta, and the American Horticultural Therapy Association. Editorial assistance was provided by Joan Stohrer, Kalamazoo, MI.

SUMMARY. This paper addresses a problem that faces all life on Earth. We, as human beings, are almost literally a cancer that is eating away from the resource base for all existing life forms, leaving behind a polluted desert in our over-consumptory wake. For other animals and plants to survive, we must face the fact that our human addictions to doing whatever feels good at the moment must be curbed.

A quarter of a million people are added daily to the Earth's surface, making even more disastrous the potential problems we face. Human beings are more a part of the Earth's problem than a solution; we have met the enemy and it is us. So what shall we do?

At every level of our individual daily personal actions, we must begin to rise to the challenge of living our lives using fewer resources. Each person, starting now, must begin to spend fewer hours with the lights turned on, encourage our children not to have children, build smaller homes, walk and bike instead of driving combustion-engine cars, take fewer trips to shopping malls, build fewer shopping malls, etc. In short, we must each of us dedicate our lives toward practicing the necessary art of living well with less.

> The white man is a stranger in the night who comes and takes from the land whatever he needs. The Earth is not his friend but his enemy, and when he has conquered it, he moves on. He kidnaps the Sky like merchandise, and his hunger eats the Earth bare leaving only a desert. Humankind has not woven the web of life. Whatever we do to the web, we do to ourselves. All things are bound together. All things connect. Whatever befalls the Earth befalls also the children of the Earth.[1]

My professional life as a psychologist has been dedicated to the exploration of the nature of humankind. I have attempted to find the truth as it relates to people and report what I have found. The words presented above I pass on to you in the spirit of Chief Sealth, a native on this continent respected as an elder among his people. The points they make are even more poignant today than they were at the time the seeds of his message were planted. We human beings are almost literally a cancer on the Earth, eating it bare and leaving only a desert in our over-consumptory wake.

When I was in the eighth grade, I was taught that we can neither create nor destroy the basic ingredients of the resources that are required to support our lives. Later I learned to refer to this statement as being basic to Nature—the first and second laws of thermodynamics. The amount of energy in the Universe has been fixed for all of eternity. It can, however, be changed in form—not destroyed, but changed. God is supposed to have spoken to Moses through a burning bush. Perhaps God speaks to us again today through the words originating from an Indian Chief who told us:

> What are human beings without animals? If all the animals ceased to exist, human beings would die of great loneliness of the spirit. For whatever happens to animals will happen soon to human beings. Continue to soil your bed and one night you suffocate in your own waste.

Near the end of 1990, *E Magazine* contained a cover story entitled "Sheer Numbers." (Hardin, 1990) In it, the point was made that population problems are chronic. What we consider news consists of sharply-focused occurrences (Hardin, 1990). What did President Clinton say today, or ex-President Bush, or some other human politician or college president or research professor? What plane crashed or where did the Earth shake and how many were killed? Extra, extra, read all about it: 263,000 people were added to Earth today. Next day, the same news. Next day, the same again, maybe a few more. Then again and again and still again. Boring—turn that thing off; I'm tired of hearing that a quarter of a million new people are added to the planet each day. Increases in population just aren't news.

Continue to soil your bed and one night you suffocate in your own waste.

As I said before, I study the behavior of humans and other animals. Recently I wrote a book. I called it *RITES OF LIFE* and said that it was about the use and misuse of Animals and Earth. I quote from its Preface:

It's a happy day. Spring is breaking on the Lake Village Co-Op Farm, where I live. I've just returned from the morning chores. The sheep and cows have been fed, and the newborn goat kids have put on their early show.

Robert Joseph, the herd bull, has had his back scratched. The chickens, pigeons, rabbits, peacocks have settled back into doing what they do all day. The horses broke through a gate and went back to the long pasture two days ago and have not been up to the barn since.

Most of the 40 people from the Co-Op have gone to school and work. They drove cars, which helped add to the millions of pounds of toxic waste we in Kalamazoo County spread out each year into our air, water and earth.

At school, the students will learn how to become an integral part of our society and thus later be able to join their fathers and mothers, as we all earn money to buy the things we think we need. Much of what we buy, of course, will end up as trash, and next year's pound-count of toxic waste will be that much higher. This fact will prompt more meetings at universities around the world by educated experts who, having received millions of dollars to research the problem, will come out with additional reports. Requests for more millions will then be made for further studies.

Trees around the world are coming down at a rate of 23 million acres per year, denuding an area the size of England—trees that provide oxygen so that we can breathe and counter the toxic wastes put out by the autos we drive to work. In our county, we are taking trees down from along our roadsides for fear that they might kill the drivers who hit them.

Soon, I will drive to work, adding to the pollution, and meet a graduate class. During the semester, they read books such as *Walden Two, Friendly Fascism, Limits to Growth, Entropy* and *When Society Becomes an Addict.* They saw movies—*Never Cry Wolf, The Animals Film*—and visited with some Amish Mennonites, ate with them, and saw how they lived. Students also came to Lake Village and stayed there with me and my extended family...and observed us as we live with all our addictions, family feuds, gates left open, recycling attempts and other issues which relate, basically, to the use and misuse of resources.

So...now I'll tell you the conclusion of the psychological research that I have been conducting with animals and our environment since 1955 and to which this book relates: Human beings (God love us) are a real problem...and unless we surrender to the other animals and watch and listen very carefully to them, as they show us how to live lives that are more in balance with the laws of nature, we will disappear off the face of the Earth sooner than later. We must, as humans, learn how to live using fewer resources and to show greater respect for all animals and all of life.

I hope that you enjoy this book, should you decide to read it. If you can bring yourself to it, however, I suggest that, between chapters, you also spend time planting trees. I know that planting is more difficult than sitting down reading and writing, but if we do not plant trees, and do not stop driving our cars so much, there are not going to be any wild doves around any more, singing...as one is doing just now down by the lake. (Ulrich, 1989)

It was about four years ago that I wrote those observations. Since then, the intensity of my concern has grown. I believe today more firmly than I did then that we humans are the Earth's cancer, her AIDS virus, her most critical illness—Mother Earth suffocating in our waste. For reasons unknown to me, we seem to produce more entropic pollution per capita than any other life form. Even with the realization of that fact, it's still hard to tell my children not to have children, although I do and to date they haven't, which I doubt is due to anything I have said. I believe that the Earth is overburdened with a destructive life form; I have met the enemy and it is me.

Well, there you have it. I have studied human beings closely and the results of my research suggest that the Earth would be better off growing trees and other plants, not more humans. I feel that these results are supported and highlighted by an NBC News report entitled *EarthWatch,* in which Dr. Paul Ehrlich dramatically narrates the global problems we now face, such as the human population explosion, species extinction and the greenhouse effect. (The video *EarthWatch* can be obtained by writing NBC Studies, New York, New York.)

As a psychologist, I feel that it is important not only to present findings based upon research but also to point the direction toward new experiments, and for me there is no longer any experiment of importance other than the real situation. And what is that? I feel the real situation is one in which all of us will sooner or later come face to face with the limits to growth that determine and, by their interactions, ultimately define what happens on this planet. These limitations are determined by population increase, agricultural production, nonrenewable resource depletion, industrial output and pollution generation.

In 1972, an international team of researchers fed data on these five factors into a global computer model and the behavior of the model was then tested under several sets of assumptions to determine alternative patterns for the future. The resulting message of that study was that the Earth's interlocking natural system of global resources could not support the rate of resource depletion that was then occurring. Thus, as humans continue to expand and grow, the results will be the same: resource depletion and the ultimate destruction of life-giving habitat will increasingly manifest itself all around us. (Meadows et al., 1972)

My exploratory travels around the world, my past 20 years living close to the earth at the Lake Village Experimental Community, and what I taste in the water, smell in the air and touch on the earth all affirm what many scientists as well as others have finally come to realize and that the spirit of Chief Sealth so dramatically proclaims:

> We must live more respectfully with regard to our resources and all our relations. We must begin to work toward living well with less.

The over-consumptory human habits that have evolved ignoring the universal truth that *"there are limits to growth"* are in no small way a direct function of the widening gap between our daily human actions and what we know as the facts of entropy. We say we should use fewer resources, but we don't do it. Exploitation perpetuated in our own lifestyle needs to be addressed. The myth that state and national well-being requires more and more resources being poured into institutional research and researchers that have repeatedly demonstrated allegiance to the ethic of "Big is beautiful," "More is better" and "Money is our savior" must be exposed so that we understand that we ourselves have become over-consumptive addicts just as Anne Wilson Schaef suggests is true for all of society. (Schaef, 1987)

Unlike many other life forms, human beings seem to be more prone to respond to short-term contingencies. If it feels good, do it again. The principles of reinforcement have shown this to be true even of non-human animals, especially when confined and forced to exist in unnatural laboratory settings. Humans, however, seem to be unbounded in their ability to over-

At Lake Village in 2010.

come limits of resource depletion displayed by other life forms who live in closer balance between the intake and output of energy and matter.

Since 1970, the human population has grown by 47%. We add one quarter of a million people to Earth daily. Agricultural production fails to keep pace as people starve in alarming numbers. Resources decline as industrial growth continues to sully the Earth with pollutants.

Neither behavior analysts nor humans as a species are Earth's saviors. We are, as the basic assumptions of the analysis of behavior implied, simply a natural, lawfully-determined part of the ongoing whole. We are no better than any other life form and above all need to become humble in this regard.

In his book *The End of Nature,* Bill McKibben laments the harm we humans have done to the Earth and pleads for its restoration. In talking of Nature's end, he suggests that it is finished as a force independent of human input.

> Whatever we once thought Nature was—wilderness, God, a simple place free from human thumbprints, or an intricate machinery sustaining life on Earth—we have now given it a kick that will change it forever. (McKibben, 1989)

An old copy of the *New Yorker* contained an advertisement from what in 1949 (the year I graduated from high school) was still the Esso Company. It summed up our century to this point: "the better you live, the more oil you use." And we first-world behavior engineers live well. The trouble is that this pattern of behaving to which we are addicted seems not to be making the planet happy. The atmosphere, the forests, the grasses and the water are all less satisfied than we are.

We need behavioral research—research that will show us how to change human behavior from actions that suggest we are more important than anything else. We must replace the Judeo-Christian ethic, which promotes the "humans are best" attitude, with new behaviors that hopefully might slow down the destruction of our Earth. We must begin to think and act more like our brother and sister plants, the trees, like the lakes, the mountains and the wind, and become sensitive to the fact that our nature is identical to the nature of the Universe. I have a friend, Rolling Thunder, who tells me that the Earth, of which we are a part, is a living organism, the body of a higher individual who has a "will" and "wants" to be well, who is at times less healthy or more healthy. He says we must treat our own bodies with respect and it's the same with the Earth—when we harm the Earth, we harm ourselves; when we harm ourselves, we harm the Earth.

Another friend I knew well, B.F. Skinner, suggested in his book *Walden Two* that we must "experiment with our own life…not just sit back in an ivory tower somewhere as if your own life weren't all mixed up in it." (Skinner, 1976)

Some years back a couple of close relatives of mine handed me a book written by Doris Janzen Longacre entitled *Living More With Less*. This book is a pattern for individual living using fewer resources. It holds a wealth of practical suggestions from the worldwide experiences of the Amish Mennonite culture of which my roots are a part. For each of us, the first step toward healing the wounds which have been inflicted upon the Earth begins right where we are at this moment. Look around you; turn something off that is using up a resource that is already needed by other animals and plants as well as by our children's children's children.

Note

1. In about 1854, on the occasion of Washington Territorial Governor Isaac Stevens' first visit to that area, a Duwamish and Suquamish Puget Sound Indian named Chief Sealth is said to have made a speech. Henry Smith, a local man, attended and took notes. More than 30 years later, in 1877, Smith was purported to have reconstructed the speech for a Seattle newspaper. Seattle historian David Buerge concluded that although Smith's text reflects at least some of what the chief had to say, the fact remains that the English version has been no doubt greatly changed over the years.

The seeds of Chief Sealth's original thoughts continue to grow, constantly finding new expression. A University of Texas professor allegedly used Smith's reconstruction for his own version, which Ted Perry is supposed to have heard at a 1970 Earth Day celebration. Perry then rewrote still another version which appeared as part of a film script produced for the Southern Baptist Radio and Television Commission. It has been expanded upon by Britain's Prince Philip and Joseph Campbell and is, in fact, the basis of a recent children's book entitled *Brother Eagle, Sister Sky: A Message from Chief Seattle*.

Wayne Suttles, an emeritus professor of anthropology at Portland State University, is said to have branded the speech as a fake. Jack Hart claims that the speech is nothing but a "…phony snippet of greeting-card blather…" (Hart, 1992), which raises questions about a trend to accept what we want to hear. In my opinion, far too much of all history has been an acceptance of what a powerful few wanted others to hear—and act upon as if it were the truth. Today, people seem to want to hear more and more about how to save the Earth. They seem to know that something has gone wrong and they are attracted to the spirit of Sealth's expanding influence. The indigenous spirit that lived respectfully with nature is touching many of today's youth. Some, like Hart, discredit that spirit by suggesting that the words don't count since Sealth didn't really say them. Others also seem frightened by our new-found interest in Nature and natives and spend great energy trying to prove that whites have invented the "good" Indian. (Clifton, 1990)

No doubt there is a little bit in each of us who perhaps would like the world to be something that it isn't. We would like it to afford us endless bounty with ever-increasing access to more and more of everything we want. I am not concerned with who really said the above words; I am concerned that we pay attention to their essence.

References

Clifton, James (1990). *The Invented Indian,* Transaction Publishers, New Brunswick, NJ.

Day, Bill (1992). *Detroit Free Press,* April 4, Detroit, Michigan.

Hardin, Garrett (1990). Sheer Numbers, *E Magazine,* November-December, pp. 40-47.

Hart, Jack (1992). Putting Words in His Mouth: Chief Seattle's Speech Raises Questions About Trend to Accept What We Want to Hear, *The Oregonian,* February 9.

Longacre, D.J. (1980). *Living More with Less,* Herald Press.

McKibben, B. (1989). *The End of Nature,* Anchor Books.

Meadows, D.H., Meadows, D.L., Randens, J., Behuens, W. (1972). *The Limits to Growth, a Report for the Club of Rome's Project on the Predicament of Mankind,* A Potomac Associates Book: Universe Books, New York.

Rifkin, J. (1980). *Entropy,* A Bantam New Age Book: Viking Press.

Schaef, A.W. (1987). *When Society Becomes an Addict,* Harper and Row.

Skinner, B.F. (1976). *Walden Two,* MacMillan Publishing Co., Inc.

Ulrich, R. (1989). *Rites of Life,* Life Giving Enterprises, Inc., P.O. Box 404, Kalamazoo, Michigan 49005-0404, pp. iii-v.

Urban Sprawl Fact: We Can't Eat Money

Reprinted from The Kalamazoo Gazette, *July 25, 2002*

The July 13 *Kalamazoo Gazette* emphasized problems surrounding "Urban Sprawl," listing suggestions from a report on land use entitled "Smarter Growth for Kalamazoo County," which stressed as one of its major features: "The need to develop a county-wide land use system."

The article named a number of groups and individuals who contend that there is no problem regarding current land use. Such denial flies in face of the fact that: No social system nor the cultural ethic it sustains can possibly survive the incredible imbalance between those who are growing food (1.5 percent) and those who are eating it.

I reside in Pavilion Township wherein the supervisor and other leaders face constant pressure, from both other governmental units and the corporate/business private sector, to facilitate economic growth more or less regardless of its consequences on farming.

Also, far too often it is the township leaders who are maligned as the ones unwilling to "give up control" in order to cooperate with county-wide "smart growth" land-use planners as if they and farmers are at fault for not halting urban sprawl. Such blame, of course, is misplaced. Every day a quarter of a million more people are added to Earth's human population, all of whom, relative to other species, bear major responsibility for lifestyles that model an exploitive overall disregard for the health of our earth, air and water, as well as growing increasingly disconnected to the fact that we can't eat money.

On Preservation of Farmland and Open Space

Unpublished letter to the Mayor of Kalamazoo, MI, Hannah McKinney, dated November 29, 2005

Dear Mayor McKinney,

I would like to discuss with you and the City Commission some thoughts concerning the preservation of farmland and open space, while at the same time strengthening the health of the city. We at Lake Village have established an open space easement, which is an agreement to avoid non-farm development. In short, we have promised to never allow urban sprawl to occur on some 200+ acres of our land. I would like to explore with the city how this effort on our part might work for the betterment of the overall community. Farms need to house farmworkers. Instead of building new houses on farmland and provoking more urban sprawl, we would like to explore the possibility of the city giving us a house where we could house people who would be willing to commute back and forth and help us with our farm chores (i.e., gardening, hay baling, fence building, barn repairs, feeding and caring for animals, etc.). Both the city and the farm could benefit. Workers could be housed downtown, and, in addition to helping on the farm, they could go to school and possibly also work at jobs in the city.

Very few small farms make enough money from their produce, crops and animals to meet expenses. Small farms are very labor intensive and must rely on outside income to keep going. In other times, people lived in the village and worked in the gardens and fields that surrounded the housing settlements. The plan I am suggesting would be similar. It would be especially educational for college students who live in town to do some work on a country farm as well as in town gardens. Far too often, people today have little or no understanding of how their food develops. This plan would help them better appreciate the food chain and the work involved. We need to draw the city, and the countryside from where our food comes, into a more cooperative relationship.

Former Mayor Jones told me that houses are available for non-profit corporations. To upgrade Lake Village Homestead is under the umbrella of a long-standing (since 1970) non-profit corporation which meets 501 c3 criteria. By helping us acquire a house in the city, our efforts to preserve farmland and open spaces would be rewarded and would help arrest urban sprawl. Such a venture could be done in cooperation with Habitat for Humanity. Farmworkers living in the city would also be able to send their children to Kalamazoo schools and thus benefit from in-city housing and the Kalamazoo Promise. As things stand now, farm children in the townships are excluded from the privilege of the "Kalamazoo Promise." We cannot move our farm business to town, i.e., raise chickens, cattle, hogs, sheep, goats, etc., there. Town people, of course, must eat, and, therefore, do benefit from farm programs and labor.

Such a plan should also involve getting faculty and administration at Western Michigan University, Kalamazoo Valley Community College and Kalamazoo College involved in helping stop urban sprawl, upgrading available houses and gain some experience in the actual production of their food.

I have spoken with Mr. Pat White, Pavilion Township Supervisor, and he is receptive toward this idea, which also reinforces inter-governmental cooperation. We might find that there are many ex-farmers (seniors) in the city who would enjoy mentoring younger people in the tradition of sound and healthy food practices. Certainly we could help with inner city garden groups.

Finally, it is extremely important at this time for the media and our educational, as well as all of our social institutions, to face the fact that **they must begin to do more to teach students something other than just words about how important healthy air, water and soil are to**

their economic survival. *Money has no value except in relation to the bounties of a healthy, more natural environment.*

I would appreciate meeting with you, discussing this plan further, and exploring what steps to take next. Members of our group will try to attend a city commission meeting soon and raise this question during public input time.

Roger Ulrich

Can Congress Dictate What We Can Eat?

Reprinted from The Kalamazoo Gazette, *November 12, 2006*

A visit with Amish friends in Indiana found them saddened by the massacre at the school in Pennsylvania. I later read how Amish neighbors had gone to the killer's parents and offered forgiveness for the shootings.

Another topic of great concern was passage by the U.S. House of Representatives of a bill now before the Senate (S 1915), proposing: "To amend the Horse Protection Act to prohibit the shipping, transporting, moving, delivering, receiving, possessing, purchasing, selling, or donation of horses and other equines to be slaughtered for human consumption and for other purposes."

The livelihood of the Amish depends on horses, especially those who breed, train, buy and sell them, sometimes even for the purpose of human consumption. If passed in the Senate, this bill will send shock waves throughout the Amish culture, as well as all other constituencies whose well-being relies upon horses.

Back in Kalamazoo, I visited the office of U.S. Rep. Fred Upton, R-St. Joseph, and learned that he had voted for passage and that U.S. Sen. Carl Levin, D-Michigan, was one of its sponsors in the Senate. In light of its potential economic damage, supporters of this bill are certain candidates for another show of Amish forgiveness.

Senate Bill (S 1915) states:
1. Horses and other equines play a vital role in the collective experience of the United States and deserve protection and compassion.

So do buffalo, bears and turkeys! Those who show disregard for their obligation to protect and show compassion for animals should be vigorously prosecuted using laws already in place. Compassion for horses is an ethic that Congress has no credentials for assuming any credibility from which to legislate unnecessary new laws. Horses transport the Amish and work their fields without resorting to the level of use of non-renewable fuels to which so many of us are beholden, and for which the United States, as a nation, wages wars to acquire.

On Sept. 7, 2006, Congress passed a bill that, if affirmed by the Senate, will put a knife in the back of the environmentally-sound practice of using horse power for transportation and work. On Oct. 20, 2006, I received a letter from Congressman Upton in which he stated: "Conservation and alternative fuels are some of the most important tools we have to reduce greenhouse gas emissions. It is important that the United States lead the world in alternative fuels as we have led the world in so many other innovations."

Too often we talk of conservation while riding in vehicles that represent the antithesis of what is said. Alternative fuels such as hay, oats and pasture are often overlooked, as are horse-powered implements and buggies used by the Amish that translate the congressman's words into an ethical reality. Passing laws that dictate the buying, selling, killing and eating of only certain animals is seemingly an infringement on citizens' rights to own, buy and sell property and, in the case of this bill, a "taking without fair compensation." It is also a denial of the way organic life naturally works.

2. Horses and other equines are domestic animals that are used primarily for recreation, pleasure and sport.

This is not true for the Amish, who most certainly did not lobby for this bill crafted by organizations whose members seem not to have the same level of appreciation for the natural cycle of life experienced daily by those who use horses to farm their land.

3. Unlike cows, pigs and many other animals, horses and other equines are not raised for the purpose of being slaughtered for human consumption.

Early humans hunted horses for consumption. Around 6,000 years ago, humans began herding and riding them. Fossilized teeth indicate that horses had bits in their mouths 500 years before the invention of the wheel, which led to horse-drawn transportation. It is true that nowadays horses are bred with the expectation of first performing other services. Nevertheless, they are still ultimately destined to become food for other life forms, as are all God's creatures. Milk cows and goats are also not initially raised for human consumption. They are raised to produce milk and other dairy products. Sheep are raised for wool. Chickens are raised to produce eggs. Each is eventually recycled into food.

What can we expect next from Washington lawmakers? Will we no longer be allowed to hunt and fish? Will attempts be made to dictate an acceptance of only so-called "natural deaths"? Will we eventually be required to embalm and preserve deceased animals in air-tight vaults as is now done with humans? If those absurdities seem simply too silly to even imagine, understand that, to many, it is not all that less absurd than the proposal the House of Representatives has placed before the Senate.

If passed, Senate Bill 1915 would make illegal the humane sacrifice of horses and their subsequent recycling into food for individuals whose customs include the privilege of choosing what they will and will not eat.

In his new book, *The Audacity of Hope,* Sen. Barack Obama states:

> "Not many Americans would feel comfortable with the government monitoring what we will eat."

It appears that we are about to find out whether or not what he says is true.

Please share your thoughts on this issue via letters to the editor or by contacting the Coalition for Natural Justice. P.O. Box 1266, Portage, Mich., 49081 or e-mail lvcnj@sbcglobal.net

Saving Our Countryside Holds the Key to Saving Our Cities

Reprinted from The Kalamazoo Gazette, *March 20, 2006*

This viewpoint concerns the fact that we must save our countryside in order to save our cities. In his book *Pagans in Our Midst,* Andre Lopez points out how: "Modern folklore finds some humor in the conflict between the country dweller and the city dweller. Almost everyone has heard country bumpkin jokes that make out country people to be somewhat stupider than city folk, a way of looking at non-city people that goes back to the earliest town people and remains today."

Lopez explains there is a reason for this conflict and invalidation:

> "Cities in fact depend for their existence upon the ability to exploit the physical resources of the countryside. That the countryside may be populated by people who do not particularly want to be exploited in no way changes the relationship of the city to the countryside. A city cannot produce all its own clothing, its own food, its own fuel, etc.... Those things must be imported from the outside. This is why cities generally function in areas other than agriculture. Cities are centers of technology and that technology must by definition be an extractive technology. It must be a technology which, whether by physical or social means, brings the goods and services which the city dwellers must have to survive".

It is important that all citizens understand the purpose of the city in this respect—cities exploit the countryside to the fullest extent allowed under the laws of physics and the laws of our judicial system. When the extractive process no longer works, when there is no longer the sustenance provided by the countryside and the natural world, the welfare of cities declines and eventually falls into dysfunctional chaos. This fact, however, is kept well-hidden by our educational institutions and the media, wherein the alleged greater importance of a "city ethic" sprawls ever farther over the countryside. For example, a Feb. 21 *Kalamazoo Gazette* editorial on the 6th State of the Downtown address, "Two lessons in downtown development," stated that one of the lessons learned by David Feehan, first Downtown Kalamazoo Inc. president and now president of the International Downtown Association, was: "Public-private partnerships are absolutely vital to making downtowns thrive." What about saying instead: Public-private partnerships are absolutely vital to making all areas of the county thrive?

When the *Gazette* says later on in the editorial that "unless people who invest in downtowns are turning a profit, downtowns will remain charity cases," shouldn't we read as well: "unless people who invest in the countrysides are turning a profit, the countrysides will remain charity cases"?

When that *Gazette* editorial says that public and private sectors "need to work in concert for the health and well-being of the urban center and larger community that it serves," should we not equally acknowledge that "we need to work in concert for the health and welfare of the countryside and the larger community that it serves"?

The following is a statement along the same line that I feel is frequently made by *Gazette* writers as well as other local leaders:

> "A strong core city with a vital and flourishing downtown support surrounding communities, thus it is in the best interest of all county residents to support the growth and development of Kalamazoo central city."

A more true and factual statement, however, should be read:

> "A strong vital countryside wherein the quality of the soil is of such richness so as to allow the production of food and other necessities that in the end, along with pure air, water and the sun, support the growth and development of everyone."

Everyone in the county needs to honor all citizens from all walks of life and from all its various places of residence. The constant emphasis upon the downtown area being the focal point for the most respect and attention must be re-evaluated if the "talk" (by various leaders in government and the private sector) that everyone's welfare is important is to be taken seriously.

Indeed only until all citizens, no matter where they reside, come to recognize and see to it that this fact becomes the "Kalamazoo Promise," to be taught to all children throughout our county and to people everywhere, will the free-fall slide into the resounding disaster of worldwide starvation stop knocking ever louder at our local door.

Living on the land is the best way to learn respect for the Earth. Alex Gibbs, son of Roger's daughter Kristan, has lived his whole life at Lake Village, where he has learned farming by living the farm life.

CHAPTER 6
Speaking Out

In Chapter 6, we see Roger as local activist, expressing his convictions in a variety of ways, through many letters to the editor and opinion pieces printed in the local *Kalamazoo Gazette* and sent to the Kalamazoo City Commissioners, and in correspondence to President Bill Clinton and to the American Civil Liberties Union. In each, Ulrich speaks from deep conviction. His "Viewpoints" have included the advocacy for the rights of Lake Village Homestead, preserving farmland, eliminating homelessness and, above all, for sharing the limited life resources we already have.

For additional articles on this subject, go to www.lakevillagehomestead.org/media

◄ *"So you want to talk about who is the boss?"*

We Must Cut Racist, Exploitive Acts

Reprinted from The Kalamazoo Gazette, *September 29, 1995*

Racism is a belief that human races have distinctive characteristics determining their respective cultures, that one's own race is superior and therefore gives it the right to rule others.

I happen to be Caucasian from Central European stock and the Amish Mennonite culture. Intellectually I don't think my culture or race is superior nor that it has the right to rule others. Yet as I recall early lessons at Mennonite Sunday schools, I would have to admit what I was taught fit the above definition for "racism." Indeed, we learned that anyone failing to confess faith in our Christian culture was in no uncertain terms a "Hell-bound inferior." Like the church, my public schooling also promoted the belief that our culture was "superior" to that of others, and eventually I graduated into a society that had in place policies of law enforcement which mightily encouraged us to pledge allegiance to such an assertion.

So I, like others, aged and barely noticed our hypocrisy in accusing South Africa of racist practices while failing to recognize the Americanized forms of the same behaviors as they apply to native peoples living among us. It somehow seems easier to recognize "racist" attitudes, usually in others, than it is to understand their source. Perhaps then this article should be entitled "Confessions of a 'Racist.'"

Racism, unfortunately, is a basic ideology in the modern world which affects every living being on Earth. Originally, as pointed out by Andre Lopez in the book *Pagans in our Midst*, it had nothing to do with race but rather the exploitation of Earth's resources and the people who held them. Today racism and exploitation go hand in hand to form the single greatest threat to all life.

During our existence on this planet we have always been creatures of the natural world, hunters and gatherers, dependent upon the forces of nature for our survival. The appearance of towns and cities, however, signaled a change (not necessarily positive) about how to best manipulate nature to human advantage.

This exploitation first started with humans taking an ever-increasing share of resources relative to other life forms. Subsequently, human verbal behavior evolved into a written tradition, which grew with the printing press into an educational technology which eventually empowered a system of justice that basically fosters a disrespect of the natural countryside and its people.

We have all heard the country bumpkin jokes in which the farmer's family is "quaint, naive and stupider" than the "sophisticate" city dweller. Today such thinking is reaffirmed in the racist stereotype promoted by the Harvard scientists who authored *The Bell Curve* and a modern racism which equates high IQ with "superiority," i.e., making a lot of money, graduating from college and being white, while low IQ is on a par with "inferiority," i.e., not making a lot of money, a welfare dropout or perhaps being a seasonal farm worker of color.

Such thinking is also promoted locally as well as internationally by various laws, codes, ordinances, treaties, etc. (and the 'officials' who enforce them), which together consistently work toward the affirmation of acts which result in ever-increasing exploitation of the natural countryside. It is a vicious, albeit subtle, system whose outcome is demeaning to natural law and to the survival of all life forms.

Indeed, it is the same system that was used against Native Americans in colonial days. An attitude that held "the only good Indian was a dead Indian" is today epitomized by the situation of Leonard Peltier, who in struggling against the exploitation of his culture and race was convicted to serve two life sentences for allegedly killing two FBI agents, in spite of insurmountable evidence that he was innocent.

History, as we know it, is the story of civilizations expanding, clashing and occasionally absorbing pre-civilized peoples. If we think globally and watch locally, we can see it occur right before our eyes as Kalamazoo expands into the surrounding countryside. With ever more oppressive new codes, people move outward onto land no longer affordable to those who have historically lived closely or more respectfully to the earth. Such people, finding it impossible to afford the costs imposed upon them by the enforcers of the mandates which have been legislated to exploit their resources, are eventually forced into a life increasingly detached from the earth. Whether we are on "well fare" (i.e., "inferior") doled out by the system or are "faring well" (i.e., "superior") as the dolers, all of us have to some degree become increasingly disconnected from the earth both physically and spiritually.

It is almost universally accepted by Western thinking that human society "progresses" by some kind of natural law from the primitive hunting and gathering culture "upwards" to civilization. It is rarely noted that there is absolutely no reason to believe that the civilization process is in any way related to an irresistible progress, nor that this alleged progress is ethically or morally positive. Certainly the current condition of the Earth's environment would argue that civilization has not had a positive effect. In fact, civilization has universally been forced upon native and natural people who live close to the earth, and the oppressors have historically been members of supposedly civilized (and in more modern times "Christian") cultures. Please note that I do not mean to imply that this culture has been "Christ-like" in spirit.

Although racism serves other purposes along the way, its consistent goal is the reduction of all resistance to the eventual loss of natural resources to modern development and the resocialization of whole peoples into roles which serve the dominant society. Although color, for many, still remains a reason for the illusion of racial superiority, it is becoming increasingly apparent that the most important basis for discrimination is the color of money—i.e., do we have enough of it to pay for the natural resources we need to survive and to support acquiring the learned wants to which we have all become addicted.

Racism at Western Michigan University has recently been a much-talked-about subject. For most Americans, regardless of color, especially those of us who find ourselves as faculty, administration, students and staff within the modern university, to be calling another person a "racist" reminds me of my father's often-stated rebuke to an intemperate allegation toward someone else..."Roger, it takes one to know one." Racism equates well with exploitation and we are at times all guilty of behaving in ways which define it.

The issues which surround racist exploitation need to be addressed. We need to work harder to understand and overcome them. A good place to begin is to look closely at our own daily actions and move diligently to live lives that follow the Golden Rule rather than "Gold Rules." We must learn to live exploiting fewer resources, remembering that the air we breathe, the water we drink and the food we eat must be kept pure in order to sustain all life, regardless of species, race or form.

Urban Sprawl and Its Effect on Farmland Continues Unabated

Reprinted from The Kalamazoo Gazette, *June 6, 2004*

The growing crisis resulting from urban sprawl and the effect it is having on farmland is a crisis with roots that are embedded in the evolution of human civilization.

Most of our existence has seen us as "non-civilized" creatures, dependent upon the forces of nature for our survival. Today, however, we live having evolved into what John Livingston refers to as the "Rogue Primate," the first truly domesticated animal and the only one that drifted away from the natural world so far as to become almost totally dependent on technical ideas in order to stay alive. This ideological, abstract dependence on ideas carried with it the ever-growing power to manipulate nature, as well as human nature, and conditioned us further to become comfortable with the arrogant greed that landed with Columbus in 1492 as an environmentally-destructive and ethno-racist culture that treated native people as inferior, uncivilized pagans.

As this ethic spread, spawning conformity to its exploitive ideology, the hunt-and-gather lifestyle was replaced by the ever-changing world of the techno-agribusiness, which supplies the goods and services that corporate rulers need to feed the continuation of our money-driven, consumer lifestyle.

No one should romanticize his or her own life as somehow being outside of this process. We are all, by nature of staying alive as, first, world humans, the main benefactors of the unmerciful exploitation of the countryside, which we facilitate to the fullest extent allowed under the laws of physics and the laws of the best system of justice that our money can buy.

I live and work on a small, cooperative farm. In all, we assume responsibility for about 300 acres within a loose confederation of 40 children and adults, 60 head of cattle, some goats and hogs, chickens, turkeys, peacocks, a donkey, horses, plus dogs, cats and other creatures that are symbolic of the small Amish-Mennonite homesteads that surrounded me during my youth.

At age 6, I was coerced into our public education system, and for the rest of my life I was entrenched in academia as a student and a teacher. In 1971, I returned to live on farmland with the vague goal of re-enacting a lifestyle reminiscent of my roots, with hope of facilitating less-exploitive social and environmental changes. While doing so, I commuted back and forth to the university where I, in a sense, stole back, via my words and writing, a money salary that originally earned its value from Earth's resources, which I then invested back into the land from which comes all our sustenance.

Throughout history there has been much talk about land use. No matter what is said, however, the results will be the same as in the past: more urban sprawl with concrete, roads, parking lots and the endless construction of more buildings to house an ever-growing human population. As each day proceeds into the next, we must remain aware of the reality that there is less of Earth's bounty to be shared by all the various forms of life that would like to stay alive.

The truth is, as Doris Longacre spelled out in her 1980 Herald Press book *Living More With Less*, that all of us will indeed have to do just that, live life each day with fewer life-supporting resources than we had the day before.

Finally, do not lose sight of the fact that these are just some more abstract ideas from still another academically-trained, publish-or-perish word merchant. More words are neither the real food we need to help us stay alive, nor are they, in fact, the unspoiled earth whereupon that food may be produced. In this regard, it would serve us all well were we each this spring to go out and plant some seeds that will grow into something that we can be nourished by, and take our children and grandchildren with us to the garden.

On Homelessness as Not Just a City Problem

Reprint of a letter dated April 14, 2006

To: Kalamazoo City Commissioners
From: Roger Ulrich and Tony Kaufman, Pavilion Township Farmers
Ref: Kathy Jessup's article: "Committee: Homelessness not just a city problem"

Kathy Jessup's article states: "A committee of city government leaders, service providers, downtown business and Kalamazoo's homeless wants to provide public-restroom access, evening meals and storage facilities for the region's homeless population and bite off small pieces of the complex problem of poverty. The group said the problem extends beyond the city's borders and requires cooperation from Kalamazoo's municipal neighbors."
When we neighbors from outlying farms come to Kalamazoo, we need public restroom access. Also, like everyone else, we need to eat and fully expect to pay for the privilege, even though our previous labor may have been responsible for the production of portions of our meal.
Now, in response to downtown retailer representative Mr. Jeff Weisman's complaint, "Why are we, just people from the city of Kalamazoo, discussing this issue today? This is not just a Kalamazoo issue—we need to spread the burden," *we would point out that he and others are not the only ones discussing it.* **Worldwide poverty, homelessness, starvation, war**

and other disastrous events are being felt by all. They are spreading their effects upon us all and they are being discussed and dealt with around the world BEYOND KALAMAZOO CITY BORDERS!

Jessup's article further stated that: "Kalamazoo City Commissioner, David Anderson, said all of Kalamazoo County's shelters are in the central city, which draws the region's homeless here and stresses the city's financial resources. Centers such as Ministry with Community and Kalamazoo Gospel Mission receive city services even though they pay no local taxes."

We would point out that Kalamazoo College, KVCC, Western Michigan University, community churches and all other non-profits in the city also receive city services even though they pay no local taxes. However, there is a big difference between the Gospel Mission and the other institutions mentioned. Gospel Mission Director, Reverend Brown, pointed out:

"We feed up to 500 people at a time and with minimum staff, and we have to have five to eight volunteers to help clean up after the meal.... If we can't get eight people out of a line of 150 to 200 people to volunteer, we cancel the meal. We have to do that about once every three months to get volunteers."

Again we point out that the food that the mission serves comes from beyond the city's borders. Many times people from the Mission have come out and worked along beside us as we did farm chores of the sort that, **in short, feed cities,** *providing food not only for the Kalamazoo city homeless, but also for all other people living in the city.*

Poverty and homelessness are, of course, not just a city problem. Yet statements such as the following persist and need serious re-evaluation and correction if misquoted.

"The city of Kalamazoo addresses the problem because other people refuse to and that is not acceptable. It's a county-wide problem, and we need to bring those people into this whole struggle. We face situations now where we're the only game in town."

"Those people," who some suggest *"as needing to be brought into the whole struggle,"* **are in it!** *We in Pavilion Township are in the struggle as a request for another 40 acres of farmland is up for rezoning to commercial in late April, albeit a much less serious issue than that of a local government decision in China to seize land from local farmers and lease it to foreign investors, and when the farmers protested 20 or more were beaten causing a 13-year-old girl to die.*

If, as it seems from the comments by some city leaders, we beyond the city's borders have behaved in ways that have caused them to feel that we are refusing to address certain social problems, an apology is perhaps in order. We insist, however, that a case can be made that non-city residents are doing the best they can to, in fact, deal with all of the problems that befall our city neighbors. We too have crime and have not yet found the murderers of the Polderman family, who lived just down the road from our farm. As do citizens from the city, we also have items stolen from our premises. We have fences wrecked by speeding cars driven by drunken drivers and trespassing hunters. There are poachers and cattle rustlers to be dealt with. And, yes, there are people needing food and housing who make promises to work in return for their meal and shelter and then refuse to keep their promise.

Reverend Mike Brown acknowledged some homeless people are banned if they are violent and that meals are occasionally canceled if they refuse to work. It is our opinion that persons at the Mission and throughout the city who behave inappropriately should expect to be dealt with in the same way they would if they were to behave similarly at a meeting of the City Commissioners.

Far too often our talking and complaining to one another and pointing fingers as to who is at fault seems to be a dysfunctional game being played. The attitude suggested by some city leaders that a trend toward thinking and talking for a living is somehow nobler than working as a productive laborer is making our problems worse. We feel shelter officials should be honored for the hard productive work they do, and it should be recognized that their endeavors go far beyond talking.

In closing, may we believe that all of us, city and country neighbors alike, must together cooperate in the search for a higher quality of life and, in so doing, never forget that all our efforts rest upon a healthy earth, pure water, clean air, the sun and the blessings of our Creator.

Keeping Fowl in the City

Reprinted from Rites of Life *by Roger Ulrich (1989), Life Giving Enterprises, Kalamazoo, MI, Ch. 8, pp. 106-109*

> *From: City of Kalamazoo, Michigan, July 30, 1982*
> *To: Mr. Roger Ulrich*
> *1013 Sutherland*
> *Kalamazoo, MI 49001*
>
> *Re: Keeping Fowl in the City*
> *Dear Mr. Ulrich:*
> This office has received a complaint that you are keeping fowl in the City of Kalamazoo in violation of Section 7-3 of the Kalamazoo City Code, which reads as follows:
>
>> "Sec. 7-3 Keeping of rabbits and poultry.
>> It shall be unlawful for any person to keep any rabbits, chickens, ducks, geese or other poultry within the limits of the city, unless the same are kept in a sanitary condition, free of offensive odors, and in an enclosed yard or coop which shall be located not less than thirty feet from the street line or any adjacent property line."

If you do not comply with this ordinance, including keeping the fowl at least thirty feet from property line, within thirty days from the date of this letter we will have no choice but to have a warrant issued for your appearance in Ninth District Court for a violation of the above section of the Kalamazoo City Code. Please see that this situation is corrected within thirty days so that this may be avoided. Very truly yours,
Don M. Schmidt, city attorney

Response:

I am in receipt of your July 30th letter and would like to share some thoughts with you.

The staff of the Behavior Research and Development Center (BRDC) at Western Michigan University is involved in all kinds of social research. We explore, among other things, human reactions to and interactions with animals. We are all a part of the study. You, the person who made the complaint, the police officer who visited (Mr. Murphy Sheaver), the children at the Sara Swickard School (who wrote us telling how they loved to see the rabbits and the chickens and the ducks in our yard), the people who walk by the lab, and myself, we are all a part of this experiment.

In the current project, to which you refer in your letter, the person making the complaint is an unknown variable. All the neighbors who surround us are known and are participating in the study—the three WMU Christian Homes to the east, the University Bookstore to the north, the neighbors to the south and the busy streets to our west.

All of us at the BRDC, animals included, try to conduct our experiments in clean conditions. We have found that the odors from our area are less offensive than are car fumes from Howard Street, the smell put out by the Upjohn Company in Portage, and the odor from the various human sewage waste spots in the city! We have an enclosed yard and a coop that is 30 feet away from any property that is not involved in the study.

Allow me to point out some further observations from our research. We have observed that sometimes attorneys forget that they too, just like everyone else, are all a part of the experiment. You work for us and we work for you. It becomes an issue of balance. In many laboratories throughout the world, behavioral science and its applied technology has evolved into something like a religion in which animals are used as sacrificial objects. At this laboratory we attempt to treat the animals with whom we live and observe as we would like to be treated.

However, to return to the main point of your correspondence, allow me to ask a question: **What situation** *needs to be corrected? Suppose someone were to arbitrarily complain about some of the noise, the pollution and the debris being generated by the construction of the new buildings going up on campus or in downtown Kalamazoo, or the odors and poisonous gases generated by your own car. Would you write a similar letter?* **What law are we breaking?** *People are being murdered, raped and drugged. People are abused, poisoned, driven to suicide, ripped off, allowed to drink, smoke and eat themselves to death. Persons in the so-called third world are starving while the land upon which they* **live** *and which could grow their food produces our coffee, tobacco, pineapples, sugar, rubber, etc. Yet the problem at hand is that someone living among us has complained about the ducks, chickens and rabbits, etc., in the yard of our lab. Well, the rabbits have all been killed by neighbor cats, dogs and cars...so the complainer need not worry about them. Four chickens and two ducks survived. We are doing our best to teach them how to be law-abiding citizens and not break the rules that humans find necessary.*

So far they are doing what we feel is good. They are making most people happy and have shown no interest in becoming attorneys or research professors and have not been observed building bombs, stealing, lying, owning or using handguns or knives, or borrowing inordinate sums of money and adding to the National Debt. The rooster does crow in the morning and sometimes in the afternoon...if you can hear him above the polluting din of traffic.

In closing, allow us to extend our best regards and invite you to keep in touch if you feel you can afford the time.

This letter is our formal response to your apparent assumption that we are not complying with Sec. 7-3 of the Kalamazoo City Code.

We feel we are in compliance and are sure that the ducks and chickens feel the same. The recently-deceased rabbits, of course, are watching this with rabbit spiritual humor, along with the spirits of two deceased friends, George Hunt and Don Hake, and all the dead Indians who we Western European types helped destroy according to various codes of ethics which were probably similar to Section 7-3 of Kalamazoo City.
Yours truly,
Roger Ulrich

Well, the truth is that many chickens, ducks, pigeons, rabbits and other animals have come and gone at our "back-yard laboratories." The Establishment backed off, and we still have domestic animals in the city, along with wild birds, squirrels, raccoons, chipmunks, plus research professors and attorneys.

No Need to Tear Down Waylee School

Reprinted from The Kalamazoo Gazette, *February 15, 2007*

Many folks are saddened by the fact that the president of the Waylee Parent Teacher Association and a few school officials are lobbying for the demolition of the very school to which they have pledged themselves as faithful servants.

Led by the Portage Public Schools board, these officials are pitching for a "yes" vote on the Feb. 27 bond proposal to extract more property tax dollars for projects they contend would result "in critical benefits" such as "academic excellence…advanced technology and competitive property values."

According to the dictionary, "Education is the act or process of imparting or acquiring general knowledge preparing oneself or others for mature life."

At Waylee Elementary School, the school principal, teachers, maintenance staff, bus drivers, recess and lunch supervisors, volunteers, parents and students have accomplished this extremely difficult process as well as any group I have ever observed during my half century as an educator.

If the Waylee School has ailments, fix them! Don't destroy it with the dysfunctional excuse that the school is too old (52 years). The State Capitol and most cherished buildings throughout Europe are older than Waylee, as are most of the homes upon which taxes would be levied to pay for the tear-down.

In short, there is no factual evidence whatsoever that succeeding in today's "fast-paced global economy" requires the destruction of the Waylee neighborhood school.

There is evidence, however, that buildings outfitted with ever increasingly complex computer technology result in children suffering attention deficit disorders, obesity, depression, loneliness, higher levels of aggression, lower creativity, atrophy of the senses, and an inability to relate to people and the world around them, as documented in Richard Louv's book *Last Child In the Woods: Saving Our Children From Nature Deficit Disorder.*

It's time we step back and carefully contemplate how we can do a better job with what resources we already have and, in so doing, save the Waylee School.

To support saving our Waylee School: Write P.O. Box 1266, Portage, MI 49081.

Time to Free Leonard Peltier

Reprinted from The Kalamazoo Gazette, *April 11, 2000*

I would like to make a statement on behalf of Leonard Peltier, who I've come to know as a friend who should be freed from prison. I am an emeritus professor at Western Michigan University and, more importantly, a small homestead farmer. My Amish Mennonite Anabaptist ancestors split from the Catholic Church in 1525, mainly as a protest against being forced into armies and against the practice of infant baptism. For these actions, they were persecuted by the church-state system of that time in ways that were eventually refined into the modern form of cultural invalidation currently faced by Leonard Peltier and others who persist in holding to a natural land-based ethic.

Reading Leonard's new book *Prison Writings* rekindled my concern for the rights of all life. I recommend his book as a document important to an American history that includes my great-grandparents leaving Europe in search of greater freedom and barely noticing the irony of having moved onto lands stolen from indigenous people. In so doing, they opened themselves to the karma of the prophecy, "What goes around comes around." Eventually, many of my family, like other farmers, lost "their" land to takeovers similar to those experienced by Native Americans.

Unlike most of my father's generation, I went on to college, progressed the ladder of academia, received my doctorate and became a university researcher and teacher blessed with the gift of academic freedom. In 1971, along with a group of friends, I returned to homesteading on a communal farm. There, we faced first-hand the ethic that places the acquisition of money above the well-being of the earth, air and water—an ethic that unless somehow altered will eventually destroy all life on Earth.

Leonard Peltier and others who make up the vast majority of our prison population are victims of this reality. Leonard's observations in *Prison Writings* and Joel Dyer's in *The Perpetual Prisoner Machine* speak out about what Indians, blacks and others in prison experience as a function of poverty-driven entrapment. We are taught to believe that slavery, concentration camps and the genocide of the American Indian Holocaust are history. They are instead ongoing realities of the ethic of environmental racism and how profits are made by fanning our fear of violent crime, which makes easier the rationalization of prison populations comprised mainly of non-whites. We then point fingers at the Germany of World War II, at South Africa, at China and others in order to deflect attention from our own human rights offenses. Guilty and fearful of admitting our role in perpetrating poverty by often taking more than our share of Earth's bounty, we foster and abet more of the same, which returns to us in the numerous forms of dysfunction daily reported in our newspapers and on television.

Leonard Peltier is to be respected for the courage and dignity he brings to his role as Gwarthee-lass (he who leads his people). Through a grant of executive clemency, the president of the United States can set Leonard free as a history-making gesture to honor the North American lands that Leonard's people represent. Such action is a needed step toward healing the wounds inflicted upon the health of Mother Earth when our ancestors came seeking their freedom.

Toward this end, I respectfully share the words of former Attorney General Ramsey Clark:

> "I want to tell you why the freedom of Leonard Peltier is so important. There are well over 200 million indigenous people on the planet.... And everywhere they (live as) the most endangered of the human species. Yet the survival of humanity depends upon their salvation. Leonard Peltier is the symbol of that struggle.... If we forget him, we forget the struggle itself.... Those who put him behind bars—and insist on keeping him there after nearly a quarter century—believe he has been consigned to the dustbin of history, along with the cause of native peoples everywhere. We must not allow that to continue."

"Humankind has not woven the web of life. What we do to the web we do ourselves. All things are bound together: all things connect."
—Chief Seattle

Eggs from Mentally Ill Chickens

Reprinted from The Kalamazoo Gazette, *April 24, 2001*

Forced molting involves withholding all food from laying hens for an average of 10-14 days so large egg industries can increase profits. The hen's body, thinking it's dying, lays more eggs. Governmental policy has been to keep food prices low by forcing farmers to get big or get out. This has set the stage for large corporations to take over the food industry and constantly look for ways to become more efficient, leading to greater pollution from large factory farm operations.

While the egg industry benefits financially, most egg eaters dine blissfully unaware that they are eating eggs from what are essentially mentally ill chickens, hundreds of thousands of which are left to die of starvation and sheer despair.

When finished as layers, their stressed-out bodies are sold to industries that recycle their remains into soups. Birds that starve are mixed with their droppings into feed for cattle, from which we get milk and beef products. The fact that animals carrying diseases, including stress-induced insanity that eventually passes into the human food chain, perhaps explains some of the increasing violence observed in our children's actions in public schools.

We need to put money into setting-up "renaissance areas" and "smart zones" for growing local, healthy food. We need to make life better for our city citizens and to help the countryside, via economic incentives, ward off land-consuming urban sprawl. We are all in this together, and in the end NIMBYs all share the same backyard.

CHAPTER 7
Guiding Philosophy

In the final chapter, we experience Roger as philosopher and artist, through a range of viewpoint articles as well as poems and dramatic pieces. While each article in this book presents a distinct aspect of his beliefs, here Roger connects how he thinks racism and exploitation of the Earth go "hand in hand to form the single greatest threat to all life." For many years, long before the current Occupy Wall Street movement began, Roger repeatedly talked about the few rich who own and control the majority of the Earth's resources (Roger cited the 5% of the world's population who appropriate 40% of the available resources). He is a vibrant advocate for the philosophy of living well with less, toward which Roger strives each day at Lake Village Homestead—from where he urges all of us to "go out and plant some seeds that will grow into something we can be nourished by."

For additional articles on this subject, go to www.lakevillagehomestead.org/media

◄ *In the early years, standing with Black, the fastest horse that has ever lived at Lake Village.*

NARRATIVES

In Search of Our Achilles Heel

Reprinted from Behavioral Analysis and Social Action, *Vol. 6, No. 2, 1988, pp. 59-61*

B. F. Skinner's insights into the problems of modern society and many of his suggested solutions are contained in the book *Walden Two*[1], which he considers his most important work[2]. In it he defrocks the pious academic contention that the solution to life's problems must necessarily come from the teaching and research of higher education.

"Some of us feel that we can eventually find the answer in teaching and research," said Professor Burris.

"In teaching, no. It's all right to stir people up, get them interested. That's better than nothing. But in the long run you're only passing the buck—if you see what I mean, sir." Rogers, the former student, paused in embarrassment.

"For heavens sake, don't apologize," replied Professor Burris. "You can't hurt me there, that's not my 'Achilles heel.'"

"What I mean sir is, you've got to do the job yourself if it's ever going to be done, not just whip somebody else up to it. Maybe in your research you are getting close to the answer. I wouldn't know."

"I'm afraid the answer is still a long way off," Burris demurred.

"Well, that's what I mean sir. It's a job for research, but not the kind you can do in a university, or a laboratory anywhere. *I mean you've got to experiment and experiment with your own life,* not just sit back in an ivory tower somewhere—as if your own life weren't all mixed up in it." Rogers stopped again.

"Perhaps this *was* my Achilles heel," said Burris.[3]

In the middle '60s, a group of researchers embarked on a project concerned with exploring various ramifications of human, animal and plant relatedness within an extended community complex. Emphasis was placed on how the human species might grow intellectually, socially and spiritually toward a greater balance with the natural laws of the Universe, while simultaneously becoming more mindful of their ethical obligations toward all of life. An additional basis for this extended project were years of studying animal and human behavior, and although not realized at the onset, for the author, an Amish Mennonite heritage. This exploration was not confined to the university laboratory, nor to books, nor to the Lake Village experimental community. Personally it took me at times to many nations and to all but one of the 50 states, often living among Native and Natural people. Most importantly, in all my quests for a better understanding of the extremely elusive nature of the subject matter, I as the explorer knew deep down that at all times I was as much the subject as I was the researcher.[4,5]

From it all, Skinner's words have proven prophetic. Efforts on the part of educators to whip somebody else up to solve our most pressing social problems have not seemed to have solved much of anything. Indeed, since its inception it often appears that classical education has contributed as much to the formulation of modern problems as to their solution.[6]

To begin with, we in higher education have taught all the public and private school teachers. They in turn have taught and influenced not only future parents but legislators, plumbers, nuclear physicists, lawyers, theologians, chemists, bankers, engineers, actors, presidents, electricians, governors, nutritionists and the school drop outs. It seems everyone has been "whipped up or out" by someone or another that the universities have influenced.[7] Even many of today's farmers feel the results of the thinking of those in higher education who taught them how to desire more dollars at the expense of respect for the Earth. The admonition to teachers to do something other than simply whip someone else up and to actually begin to experiment with one's own life is similar to the plea *"physician heal thyself."*[8]

The observation that the acquisition of answers to our pressing social problems is a job for research, but not the kind you can do in a university or in a laboratory anywhere, is of course nothing new to traditional Native American Indian spiritual leaders.[9,10] To them, man's inner nature is identical with the nature of the Universe and thus man learns about his own nature from nature herself. The technological and materialistic path of contemporary Western society is the most unnatural way of life man has ever tried. Traditional Indians see the people of this society as being far removed from the trees, the birds, the insects, the animals, the growing plants and the weather. The "educated person" of modern society is, therefore, the least in touch with his or her own inner nature. Unnatural things are so commonplace to the modern mind that it is little wonder natural things seem strange and difficult to face. Indeed, for many modern Americans, important perennial truths seem like new revelations.

Experiments do not cause things to happen. Events are caused by their natural causes. University classrooms and laboratories do not represent a natural setting. There is no experiment other than the real situation. The university is not the real situation. Yet it is often touted to be the place where society should turn to reach the answers that will solve the very real problems we find ourselves facing.[11] If this is true, from where then do the problems come with which we are today faced, i.e., poisoned air, earth and water plus the endless array of human social concerns which seem to grow daily in severity?[12,13]

In spite of Skinner's words in *Walden Two* and current worldwide events as seen so graphically on the nightly TV news, university leaders continue to insist that they need more of everything, especially money. Former students, well taught by the examples set by their teachers, not surprisingly go along with their demands. Indeed, we have taught our students to behave in ways which perpetuate the myths created by the very institution that was supposed to be above perpetuating myths. Since there is no experiment other than the real situation, let us turn now and take a close look at the "real university situation."

All universities strive toward becoming better places to learn and places of which we can be truly proud. Places where students are said to be the central concern. Classic speeches by university leaders tell over and over of our attempts to maintain an environment in which students can grow, not only intellectually but also socially as human beings sensitive to the needs of others, and mindful of their ethical obligations towards the society in which we all live. Indeed, I would suggest that no one has ever heard a university spokesperson say publically that students were not the central concern.

Yet, in spite of such rhetoric by university leaders all over the world, what has grown the most at universities almost everywhere are the number and the size of buildings, the number of administrative managers and back-up staff, and the number of dollars that they are paid.[14] At the same time, students struggle to pay ever-increasing tuition, while the money available for the jobs they seek as they go through school has grown very little if at all. All over the nation, taxes go up and dollars are passed throughout universities and the other institutions they have spawned. In the name of teaching and research, millions and millions of dollars for construction of buildings is promised by legislators and executives who were taught that the answers to social problems lie within the university. University leaders are chosen for their ability to convince governors, legislators and the judiciary that we need more and more money to teach our students the skills required for success in our society, as well as those necessary to become ethically-responsible persons with global insights and concerns. University boards of control, who pick the university leaders who claim to teach students the ethics of not taking more than their fair share, are appointed by the very governors and legislators who the university leaders will soon lobby for more money. *Missing more and more often from this circle of invested and conflicting interests are those persons close to the Earth, from which all resources sooner or later come.* People of the earth, who in the end are accountable to the rules of Mother Nature, are seldom consulted in any meaningful way as to how the Earth's resources are to be divided. Often people of the earth are looked upon as *"dumb farmers,"* not bright enough to keep from going broke, and with little understanding from the highly verbal salesman within the university establishment. For example, Western Michigan University has approximately 18,100 full-time-equivalent students. At this university, there are approximately 445 separate classrooms and laboratories presently available containing 20,501 seats. That is more than one seat per person. That is to say, every student could be seated at the same moment in a class or lab with 2,401 seats left over. There are 36 additional conference rooms not used for classes, averaging 20 seats per room, adding 720 seats. We have 460,000 sq. feet of offices for faculty and administrators and other staff with an average of four chairs per office, which adds still more seats. There are restaurants, snack bars, lounges, gymnasiums and field houses all full of chairs. There are auditoriums for plays and musical events. There are numerous dorms. There are libraries and chapels. There are fitness rooms and, of course, the many buildings for the purpose of heating and maintaining all the others. The number of seats grows as the list goes on. When the total square footage of heated and air conditioned space at Western Michigan University is divided by the number of students, the resulting amount of space per student is 322 square feet. A 20' X 15' exclusive suite for every student.

Good education was, someone once said, Mark Hopkins sitting on one end of a log with a student sitting on the other. We are now faced with 322 square feet per student with wall-to-wall seats, much of which of course is being used by the administrative managers whose names and faces many students will never know or see, because they are out convincing politicians and the public that they must buy another building or else our educational system will fold.

We have come a long way…but in which direction? Although I might not agree that sitting anywhere constitutes a sane situation for good education, I have observed that buildings and administration were not stressed as important in that famous Hopkins student formula for excellence in education. Universities continue to construct millions of dollars worth of additional buildings while buildings already in existence are underused and under cared for. *As such, are our buildings not anything more than monuments, mocking our rhetoric to work toward teaching students to be sensitive to the ethical obligations toward the greater society?* In the United States, 5% of the world's population lives on 40% of the world's resources. We live in a world where many of our brothers and sisters are starving with no place for shelter. Meanwhile, back at the citadels of our culture's centers of education, we are told that new buildings are cheaper than repairing the old in order that they meet CODE. Few, however, stop to ask, whose code? One wonders if the world's great religious centers, such as St. Peter's in Rome, the Notre Dame in Paris or Westminster Abbey in London, or many of the buildings of the world known as great educational centers, such as Oxford and Cambridge in England, would meet the codes often set in the United States as necessary for good learning to take place. There are still those, however, who understand that much can still be learned about the art of survival outside, or in a barn.[15] My son, who managed to graduate from Western Michigan University with honors and who found he could sometimes receive class credit for courses outside and away from the usual classroom (including five months of hitchhiking around the United States), confesses that those days supplied some of the greatest learning experiences he ever had—including time spent in a Mexican jail.

University education has not provided us with a problem-free society in spite of all the years it has been around. Indeed, it is possible that university education has promoted human greed as much as it has any other known human trait. Not only are members of the human community in jeopardy, but countless other species of plants and animals are close to extinction because of what humans teach.

Skinner said in *Walden Two* that it was time to experiment with our own lives and suggested that small can be beautiful, as the late E.F. Schumacher proclaimed.[16] The prestigious Club of Rome, with the help of M.I.T. computers, told us in no uncertain terms that there are limits to growth.[17] The experimental analysis of behavior has clearly shown that it is not the quantity of goods that counts but the contingent relation between goods and behavior. That is why (as Skinner has noted) to the amazement of American tourists, there are people in the world who are happier than we are, while possessing far less. Yet we in the "ivory towers of truth" continue to teach by the examples we set that we must constantly grow larger and get more.

When one is living close to the earth it becomes apparent that not every year is there a 5-8% increase in the amount harvested. When we exhort Mother Earth to greater and greater feats of production, whether it be from farming marginal lands or pumping chemicals into the earth and the animals we wish to exploit, one notes that other problems occur. For example: (1) top soil lost from overgrazing or over farming; (2) the lowering water table from the draining off of wetlands; (3) sickness among humans related to what we have fed our animals to make them grow bigger and faster; (4) forests cut indiscriminately and dying from acid rain; (5) tropical deforestation occurring at increasingly alarming rates. All this means that sooner or later we must face the fact that we cannot always expect more and bigger of everything. Indeed, out of our greed for higher production, university-trained farm advisors and engineers have led us into landscaping the countryside so it acts like city roof tops, which channel rain water to run soil-laden to the sea. In short, we cannot expect to always grow more crops or achieve greater salaries or build bigger buildings or, above all, finance bigger armies to guard the material results of our greed![18]

Inflation is said to be the most serious problem in the world today.[19] It has been defined not ineptly as spending more than one has. Other leaders point out that bigger amounts of expenditures now for whatever the purpose mean less for tomorrow's children. Squirrels don't hoard 5-8% more nuts each new year than the last, and robins don't build ever increasingly elaborate and expensive nests. They, like other creatures except for humans, live in balance with nature and are not in debt for their over-consumptive actions. If universities are ever to become a better place for students to learn, it will necessitate teachers truly beginning to experiment with their own lives and realizing that we can't whip others up to look and behave according to the truth, while we sit back in our expensive ivory towers, built on borrowed and inflated money, as if, somehow, our own lives were not mixed up with the truths we claim to reveal.

Today it is common to see people from other parts of the world become increasingly hostile to the USA, where we have yearly taken 40% of their goods to fulfill the greed of the 5% of us living here. History has shown over and over that natural laws eventually will no longer tolerate inequities of the sort perpetuated by us when we build new unnecessary buildings in the very face of people around the world sleeping homeless and starving in the streets and in the countryside. When, for whatever reason, one species finds itself so far out of balance with nature that it no longer fits into life's web, it faces extinction.

How much suffering can people of wisdom impose upon the yet unborn? We sit in our air-conditioned ivory towers with God only knows how many chairs per student speaking about creating an environment in which these same students will become sensitive to the needs of others. Those who through their own daily actions continue to act as if their own lives are somehow above the laws of nature will soon find that their examples of daily conduct are more powerful than the teaching and research they claim is so important for directing the actions of future generations.

"Well, that's what I mean sir. It's a job for research but not the kind you can do in a university, or in a laboratory anywhere. I mean you've got to experiment, and experiment with your own life, not just sit back in an ivory tower somewhere as if your own life weren't all mixed up in it."[20]

Perhaps, as Professor Skinner mused, this is our Achilles heel.

References

1. Skinner, B.F. (1948) *Walden Two*, Macmillan, New York.
2. Ulrich, R., Skinner, B.F. on The Dick Cavett Show (1969), New York, New York.
3. Skinner, B.F. (1948) *Walden Two*, Macmillan, New York, pp. 4-5.
4. Ulrich, R. (1973) *Toward Experimental Living*, Behavior Modification Monographs Vol. 11, No. 1, Behavior Development Corporation, Kalamazoo, Michigan.
5. Ulrich, R. (1975) *Toward Experimental Living, Phase II, "Have you ever heard of a man named Frazier, sir?"* In: Ramp E. and Semb, G. (Eds.) *Behavioral Analysis; Areas of Research and Application*. Prentice Hall, Inc., Englewood Cliffs, New Jersey.
6. Sanford, Nevitt (1962). *The American College; the Psychological and Social Interpretation of the Higher Learning*, John Wiley and Sons, New York.

7. Ehlers, H. and Lee G.C. (1964) *Crucial Issues in Education.* Holt, Rinehart and Winston, New York.

8. Ulrich, R. (1981) *I Am Human Conflict and Aggression,* Behaviorists for Social Action Journal Vol. III, Number 3.

9. Boyd, D. (1974) *Rolling Thunder,* Dell Publishing Company, Inc., New York, New York.

10. Deloria, V., Jr. (1974) *We Talk, You Listen,* Dell Publishing Company, Inc., New York, New York.

11. Curts, Merle (1963)*The Social Ideas of American Educators,* American Historical Association Commission on the Social Studies, Littefield, Adams and Co., Paterson. NJ.

12. Ulrich, R. (1984) No "Savior" Institution, *Kalamazoo News: The Independent Weekly,* October 26.

13. Hemenway, D. (1984) *Living Lovingly on the Earth,* The International Pernacuttual Seed Yearbook, P.O. Box 202, Orange, MA 01264, USA.

14. W.M.U. Chapter AAUP Bulletins, 1977 to present.

15. *Ibid,* pp. 8, 9.

16. Schumacher, E.F. (1973) *Small Is Beautiful,* Harper Torchbooks, New York.

17. Meadows, D.H. Randers, J. Behrens W.W. (1972) *The Limits to Growth:* A Report for the Club of Rome's Project on the Predicament of Mankind, Universe Books, New York.

18. Gross, B. (1980) *Friendly Fascism: The New Face of Power in America,* South End Press, Boston.

19. Skinner, B.F. (1948) *Walden Two*, Macmillan.

20. *Ibid,* p. 19.

The Road to Cuba: An Essay on Some Predicaments of Modern Civilization

Introduction

B.F. Skinner once implied that his book *Walden Two* (1948) was his most important work. "It's in *Walden Two*. It's all in *Walden Two*," he said in reply to a question asked him near the end of a session on the Dick Cavett TV Show (1969). In it he said that "the answers to our most pressing social problems are a long way off and although their discovery is a job for research it is not the type of research you can do in a university or a laboratory." *"You have got to experiment with your own life,* not just sit back in an ivory tower somewhere…as if your own life weren't all mixed up in it." (1948).

Some ten years ago Rolling Thunder, a Shoshone Indian leader, offered to me the idea that experiments do not make things happen. Events are caused by their natural causes. "There is no experiment other than a real life situation." (Boyd, 1974) He then gave an example of the predicament of a contemporary "science" which is often too "scientific" to include all the conditions of the real situation. The example was a story of making water into medicine and it makes very clear the necessity of being truthful about and willing to consider ALL variables of a situation, just as scientists have stressed albeit often ignored.

The example goes like this. An Indian who wants to cure a fever may be caught with no medicine. That being the case, he takes a glass of water and prays over it. He offers his prayer in the morning when the sun is coming up, when the vibrations on the Earth, which the Indians call the Great Spirit's power, are the strongest. The Great Spirit's power may be just another name for what behaviorists call the natural conditions. Whatever the name Indians, scientists and farmers know that time of day to be strongest in terms of bringing forth new life. The Indian who knows and believes this lets the rays of the sun hit his water and in doing so makes it into medicine. Rolling Thunder insists that contemporary science, as most accept it, can make no valid test of such American Indian medicine if it refuses or cannot by its own assumptions include all conditions of the real situation. In this case, the real situation includes the need for a result, which is that the Indian is sick with a fever and he needs the result of being well.

The Indian also believes that the need and his ritual will bring that result. This Indian ritual, I believe, is just a ritual different than the one of going to a doctor, trained in the European tradition, for a pill. (Behavior) = (Function) (Environment) given (Existing Conditions).

B (Indian, I am well) = (F) E (Water and Sun) given K (all the conditions of the Indian's real situation)

B (White man, I am well) = (F) E (Water and Pill) given K (all the conditions of the White man's real situation)

In the Indian's case, it includes a certain attitude toward the sun and the Earth and nature. It includes a certain viewpoint about the relationship between the sun, the Earth, the person being healed and the glass of water. Finally, it includes a feeling about the conscious presence of the Great Spirit. Absent from the real situation are skepticism and judgment, and it is exactly the same type of thing which occurs and is present when we from a different culture take our medicine and perform simply a different ritual.

The final understanding of phenomena that are outside the university research laboratory involves research and explanation that simply cannot be accomplished in a university laboratory. I am aware that many scientists would be embarrassed to investigate American Indian medicine if they should be required to work using what they might refer to as mentalistic or non-scientific assumptions. Yet I believe they are the assumptions that exist when such things as the above examples happen (i.e., both Indians and whites being made to feel better given the conditions present during their healing rituals).

I too have been a bit embarrassed at times to admit to having explored various things. Don't look there! Don't read that! Don't touch that! Don't go to Cuba! (Many people said that I should not travel to Cuba, including the U.S. State Department.) Yet there is no way to find out things about Indian medicine or what is happening on the road to Cuba without exploring the real situation. Thus you visit and live with Indians and you travel to Cuba.

Method

On December 26, 1986, Carole Ulrich, Paul Doonhague and I, driving in a 1972 open road Ford van, left Kalamazoo for Key West. Thirty-six hours later we read a sign with an arrow pointing south "90 miles to Cuba." I spent a week visiting with friends and mental health officials, whose greatest concern seemed to be how to handle problems related to what I would call a lack of temperance. I stayed with Marshall Wolfe and Dottie Marine. Marshall is the director of mental health in the South Keys, and Dottie teaches courses related to issues of drug abuse in the public schools. One night I went to the hospital with Marshall, who was on call. When we arrived we found a young man from Ohio tied down in a straightjacket. Marshall introduced me to everyone as Dr. Ulrich from Western Michigan University in Kalamazoo, Michigan. The young man was in a word acting crazy. Above all he was being disrespectful to everyone in sight. Strong people of good will were doing their best to treat him with kindness. He asked to go to the toilet. When they brought the bedpan, he urinated on the nurse. I asked him if he had taken crack. He paused in his thrashing and said, "Fuck no, man, what do you think I am, stupid?" I called Marshall aside. I said, "The kid needs to have his ass kicked." Marsh agreed and said sooner or later he will get his ass kicked. Somewhere, sometime in life he will get just that. He will be hurt, but not from any of the people here at this hospital at this time because we all want to keep our jobs, and we all know we can't use physical force. We were taught that in Behavior Modification classes at the university and it has become the law.

On the way home after having given the young man other drugs that had a different effect, we discussed the fact that because all illegal drugs are lumped into the same category, distilled cocaine has become for many the drug of choice, not because of consumers, but rather because sellers can smuggle it into the United States easier than they can bales of marijuana. Thus some humans who care nothing about anything other than the monetary profits they gain push a substance that can be a deadly poison simply because it is easier to get across borders. Alcohol is another potentially deadly poison even though it is now legal and we serve it almost everywhere, including most of our university functions. At one time in history, the United States government used it to disrupt and make easier the job of genocide among Native Americans. It had been discovered that Indians could be easily addicted to alcohol, and taking the land from people poisoned beyond comprehension of what was happening to them made the theft simpler. Our ancestral white leaders also knew that smallpox virus among Indians would cause sudden death and thus gave Indians gifts of smallpox-infected blankets. Events like those were of the nature which gave rise to the Indian saying, "Never trust the water downstream from a white man" (Boyd, 1974).

On my travels I observed other predicaments besides children on drugs which fit into the prediction by the Hopi Indians concerning the disasters which are going to one day befall this country. Disasters that are the delayed effects caused by early actions taken by the people who first invaded this country. This story by the Indian Medicine Man Mad Bear Anderson describes a basis for such feelings.

> "The principle of cause and effect is at work everywhere, and somebody has to receive the results of everybody's doings. Every sentence or thought or act has an effect on somebody. If someone has a destructive thought or wish, it has to have an effect on someone. If it doesn't work on someone else, it works back on the person who created it. Of course, in the end everyone gets his own earnings and accounts for his own debts; but just like money, it can go around and around and involve many people, and it can get very complicated. The purpose of good medicine is to make it simple. There's no need to create any opposing destructive force; that only makes more negative energy and more results and more problems. If you have a sense of opposition—that is, if you feel contempt for others—you're in a perfect position to receive their contempt. The idea is to not be a receiver.

You people have such anger and fear and contempt for your so-called criminals that your crime rate goes up and up. Your society has a high crime rate because it is in a perfect position to receive crime. You should be working with these people, not in opposition to them. The idea is to have contempt for crime, not for people. It's a mistake to think of any group or person as an opponent, because when you do, that's what the group or person will become. It's more useful to think of every other person as another you, to think of every individual as a representative of the Universe. Every person is plugged into the whole works. Nobody is outside it or affects it any less than anyone else. Every person is a model of life, so the true nature of a person is the nature of life. I don't care how low you fall or how high you climb, economically or academically or anything else, you still represent the whole thing. Even the worst criminal in life imprisonment sitting in his cell—the center of him is the same seed, the seed of the whole creation" (Boyd, 1974).

We as white people once poisoned the Indians with alcohol. Now crack distilled from an herb, the coca plant, is causing a very big problem in communities all over the United States. Could it be that the laws of payback are in operation? The Indians in Peru chew the coca leaf. They do it with temperance. In the U.S. many people of all ages are addicted to different forms of the coca leaf distilled into cocaine and crack. Given that our children, as well as adults, can't handle cocaine, we are financing expeditions to go and destroy the coca plant fields and the distilleries in South America. The equivalent to that would be for Peruvian Indians to proclaim, "Our children are being destroyed by alcohol and we are coming to the United States with armed forces to destroy the Hiram Walker distillery in Peoria, Illinois, and to burn the corn crops around the state so you can't make any more corn liquor, which addicts our people." What goes around comes around, was an old hippy saying of the 60's. What we have done unto the Earth will be done unto us. Certainly the events in South Florida Keys, including the young man in the hospital and many other events on the road to and from Cuba, seemed to be related. The United States and our allies send guns to Central and South America so that one group of people (called freedom fighters or Contras) can kill other groups of people (called leftists or Sandinistas), who are supplied with guns from Russia. The problems of these third world countries are rapidly becoming our problems as more and more killing is occurring right here at home. Someday such problems, generated by racism and exploitation, will become even more apparently ours as well as a problem of the third world. The politicizing of the issue of which drug of choice, when used, will be tolerated, and which will be shown zero tolerance to the extent of allowing the power elite to abandon all civil liberties for anyone who's even suspected of using a non-sanctioned substance, could come to produce another Holocaust where the new Jew is simply someone who is suspected of smoking a joint and thus their property confiscated and the new victim—destroyed, whether shot by a SWAT team or shipped to a concentration camp called prison (Gross 1980).

Results: Part I

After a week in Key West I drove to Miami. As I mentioned before, many people had counseled that I not go to Cuba. So I talked to as many Cubans as I could. One man in particular convinced me that I should go. He was from Cuba and had left only a year earlier. He gave me his family's address and a book to give his brother. The trip was on. Although many more miles had to be traveled to reach there after having been only 90 miles away in Key West. Methodology blends into results and vice versa.

Driving the rest of December and January, I moved across the southern part of the U.S. toward San Diego, where I again found that much of what we learn in life is not what we expected. At the University of San Diego, I met with Drs. Don Neuman, Eve Segal and other friends who provided me with introductions to people from whom new directions in my thinking were to evolve. My interest in American Indian customs led to the opportunity for another almost totally novel real situation experience. In San Diego I thought that I was going to meet an Indian Medicine Man who read "auras". Instead, I ended up at the home of John Lawrence, a white man who had studied with Yogananda, the author of the book *Autobiography of a Yogi*. Let it suffice to say that the results of that meeting again reinforced an already strong belief that we are all participants in an ongoing eternity of events in which our earthly experiences are transcended by an ongoing consciousness or collective unconsciousness described by Jung, which seems to build upon the things we learn and do during our Earth life.

Let me also say at this point that the complexity of the spiritual-material relationship is for me so complex as to be a "Great Mystery," which is the Sioux Indians' name for God. As a scientist I learned that behavior was lawful and determined and occurred according to the natural conditions. Those natural conditions have been often just as hard for me to understand as is what many call God or any other unknowable. So for the moment then I simply pause and pay respect to the Great Mystery and return to the main theme of this paper, which has to do with earthly things that occurred on the road to Cuba.

Results: Part II

In February and March of 1987, after spending time in various parts of Mexico traveling and living out of the Open Road van with my daughter, Traci, the two of us finally flew to Cuba and went through customs. We were free to move about. We saw the usual tourist sites, Old Havana, the Tropicana Night Club, the fort, monuments to the revolution, etc. We walked the streets and rode tour buses, swam in the hotel pool and talked with people from all over the world. We rented a Russian car and, with the brother of my new found Cuban Key West friend, drove out into the countryside and found the family of Benito Rodrigues, who lives at the Lake Village Commune (Dune, 1986), which started years back as a *Walden Two*-type experiment. In a little small-roomed house, with no glass in the windows, with wooden table and chairs, nestled among vegetation hiding it from the road, we were met by a tall dark man, Benito's brother—and his small brunette wife, who within seconds showed Traci and I down-home family love and affection. They were glad to see us, and from then on things got better as we met the rest of the family. In short, we were among people of the earth, we were the campesinos of Cuba! People who live in a land where the realities of life are in many ways harsh. There is not much on store shelves and the work is hard. There is little of anything wasted. They live closer in balance with the Earth than do many others.

Back in Havana there were the people of the city interacting with all the things cities have to offer. There were the diplomats from Russia, wrestlers from East Germany, there was the British visitor, the tourists from Canada, the two Americans who were from San Francisco, one a journalist and the other was her cowboy husband, both with State

Department clearance. There was the tour group with whom we were loosely associated from Mexico, with an Argentina family thrown in. All were exploring like us. There was the visit to the university, and to the monument of the revolution, all in this once U.S.-like city now drab and a bit unkempt with seemingly less for the native Cubans and more in the hotel stores for foreigners. The currency of exchange for all foreigners was the U.S. dollar. Everywhere there seemed to be the influence of Fidel and the memory of Che, and most of all the statues of Jose Marti which proclaim that the Cuban revolution was a popular, poor people's revolution. There was no outward feeling of any strong Russian influence. Cuba remained Cuban. I've been to Russia and watched Russians dance…. The Cubans didn't dance like Russians. Cuba is materially poor by first world standards. Benito's family showed me a richness of spirit that remains strongest, it seems to me, wherever I go among people who are close to the earth. Thus the people of the countryside in Cuba seemed to be strong and in many ways spiritually wealthy in spite of poverty and the deemphasizing of the organized Catholic Christian Church. A major emphasis in Cuba's international policy is probably best contained in the words of one of Castro's speeches: "The debt is not only unpayable but also uncollectible" (1985).

> "The initiative for the future is now in the hands of third world countries…confronted by a greedy, selfish insatiable world we hold a powerful weapon: to unite behind this banner and impose the establishment of the new International Economic Order."

In short, don't pay back any so-called debts, you owe nothing to the banks. What was taken from the Cuban country was taken with the force of greed…very similar to how the U.S.A. was taken from the Indians and how the Earth has been taken from other life forms by humans. In the end, it is the Earth to whom we all owe debts…not banks, not governments (Ulrich 1984). Castro's Cuban thinking is further exemplified in another speech "This is the Battle for Latin America's Real Independence." (1985)

> We, the Latin American peoples, were the ones who financed European capitalism. The gold, the silver and everything else that came out of this part of the world—produced by the sweat and blood

of the Indians, the black slaves from Africa and the mestizos—served to finance capitalism...we have been systematically plundered for centuries. That's why I say that it is economically, politically, morally, and legally impossible from all those angles to pay back the first world bankers. One of the historic reasons is that we were plundered; they are our debtors now, even in the economic sense of the word. It is Cuba who has taken up the cause over the impossibility of paying back the so-called debt allegedly owed by third world countries. Why have we taken it up? Because nobody else did. We didn't do this because we want the glory or the prestige—far from it. No real revolutionary is interested in those things. Marti said that all the glory in the world fits in a kernel of corn; that was one of the first things we learned. And a kernel of corn is pretty small.... This movement has gathered momentum; it's like a snowball that's rolling onward with unstoppable force, supported by the law of gravity—not of the Earth, but of a planet with much more volume, proportional to the immense debt that is weighing on us. Therefore, it is a snowball that is increasing its velocity as it rolls and is growing, growing and growing. By now, nobody can stop it. This is the truth. Everybody knows this (Castro, 1985).

I would respond that not everybody believes it, especially here in the United States, where almost without realizing it we too have become debtors to banks. In Cuba I found evidence supporting the beginning of a new world view that I believe is in the process of replacing the Newtonian Materialistic world machine as the organizing frame of history which will eventually overthrow all our current survival methods, which have placed their faith in a *No Limits To Growth* philosophy. The results of this experiment on the road to Cuba showed that the entropy law is more rapidly taking over as a fact of earthly life than any of the world's leaders seem to realize. The entropy law is the second law of thermodynamics. The first states that matter and energy in the universe is constant that it cannot be created or destroyed. The second law states that matter and energy can only be changed in one direction, that is from usable to unusable, or from available to unavailable or from ordered to disordered (Rifkin, 1981). There are limits to growth in spite of what politicians and other alleged leaders would have us think. We cannot deficit spend forever as we raise salaries either as a nation or as academic institutions! Paper money and non-renewable resources are more related than is often realized. Misspend one you misspend the other.

Francis Bacon was presented along with Rene Descartes and Isaac Newton by my graduate school teachers as early prophets for the experimental analysis of behavior. Bacon wasn't into sitting around and contemplating nature. He counseled that we "experiment to relieve and benefit the conditions of man and find a method for controlling nature." Bacon did not see the real situation as an experiment. So taken for the time with this approach, I became a part of a group dedicated to the *control of human behavior* and often did not consider the *limits to technological and all other growth*.

I once asked Israel Goldiamond, my main graduate school mentor, *why* something occurred. When the dust settled, I had learned that such metaphysical questions died with the Greeks except perhaps for Mennonites, who hadn't yet learned that a science of learning must be committed to the *how* of things: that was the jist of his words. Skinner, I always felt, seemed fond of Descartes, who was even more global in his presumptuousness than Watson, whose behaviorism allowed him to think he could create at will (Indian Chiefs). Descartes reduced all quality to quantity. Then told the "civilized" world that only space and location mattered. "Give me extension and motion," he said, "and I will construct the universe" (1981). John Locke (another hero for those who would place nurture over nature) emphasized bringing the workings of government and society in line with the world machine concept. Locke became the philosopher of unlimited expansion and that there was no end to material abundance. Adam Smith did the same with economics. Bacon, Descartes, Newton, Locke and Smith were the great popularizers of the mechanical world view of mathematics, science and technology. The mechanical view of materialism and progress, where people's needs and aspirations, their dreams, hopes and desires, all became slaves to the pursuit of *material self-interest and growth!*

Results: Part III

The results obtained from this study revealed more than anything else that Cuba and the road to and from it are strewn with the debris of such purely quantitative thinking, and Traci and I, as we drove and flew on our experimental quest for knowledge from real situations, were a part of a growing world catastrophe, the rapid depletion of the Earth's available resources. The huts and the so-called poor people along the back roads of the Earth are testimony to this fact, although it does not correlate to richness of spirit. The total energy content of the Universe is constant and the total entropy is continually increasing. Every action we human beings take in this would either speed up or slow down the entropy process. By the way we live we affect the speed by which the world's available energy is dissipated, and at the point of such thinking is where science joins metaphysics and ethics. Where *how* we do it (*science*) joins why we ought to (*ethics*) and the very basis of knowledge itself.

We are all on the road to Cuba, where people have far less materially than we do here. It winds through Haiti, where my cousin's children do relief work for the Mennonite Church and where the countryside is almost totally denuded, as if humans were locusts who have devoured everything in sight and are slowly starving. It snakes its way into the deserts of Ethiopia, from where we all saw the starving faces that TV crews from first world countries brought into our comfortable living rooms as first world pop heroes sang the background music "We Are The World." One and a half billion folks go to bed malnourished each night, while we here in the U.S. claim American agriculture leads the world in agricultural efficiency. American agriculture which in truth is the most inefficient form of farming ever devised. In Illinois my Mennonite uncles and cousins use 2,790 calories to produce just one can of corn, containing 270 calories. When Benito's brother knew we could stay for dinner, he went out with his dog who ran down a chicken which, added to a lot of rice, was our evening meal. One Cuban farmer with an ox and plow produced a more efficient yield per *resource energy unit expended* than our giant mechanized agrifarms could every dream of doing.

In addition to American agriculture, we thank God for American education and raise tuition for those who wish to receive it, not telling them that the history of human mental development has been a history of removing the human mind farther away from the reality of the world we live in. The current computer and microchip "revolution" is a case in point. Its advocates (i.e., salesmen) argue that it is another example of how more can be done with less. One day, it is forecast, we will be able to transmit virtually all the knowledge known to humanity and will not have to leave home to receive it; but as Diether Haenicke (1988) points out, it does not help if you can transmit a message with even greater speed or lesser cost if you do not know how to formulate the message. So when universities join in communications networks so researchers can collaborate with colleagues at other universities by sending and receiving experimental data in a matter of seconds, we publish it as positive news instead of declaring a day of mourning. *The general effect of the computer revolution has been to dramatically increase the overall entropy of the world.* Today there are millions and millions of computers pervading every facet of life, computers which promote the use of nonrenewable resources faster than the tongue can tell of a new destructive missile being built.

Discussion

By way of final discussion, it should be clearly stated that any personal illusions of being more a part of the solution as opposed to being part and parcel of the problem is always, somewhat sadly, short lived. I suppose we would all like to at times be a savior. For me Jesus was always painted in heroic colors as I was "christianed up" in what often seemed like sinful ways. The question now is one of balance. Can we do more with less natural resources being destroyed? Is there anything much we human junkies can do to stop our overconsumption? *Or are we, after all, the Earth's cancer?* Millions of people were starving to death as Traci and I rode the plane out of a smog-clogged Mexico City toward Havana and were still hungry as we returned. Are we as human beings in fact held hostage to our own technology? TV's, radios, planes, computer terminals, cars, phones, duplicators, all contributing to the speeding up of the depletion of Earth's resources. Decisions become harder to make as the world becomes more confusing to both ourselves and our children. We are being bogged down with information overload, which is a new clinical phrase which seems to sit somewhere near the entropy law and in the lives of millions of Americans being treated for more mental illnesses today than ever in the history of humankind.

The road to Cuba never ends…and the word entropy is just a modern term written in a slightly different form from the words of the oral tradition of Indians who tried to tell us the same message when we came and invaded this land. Solving the resource-depletion problem is what Granddad was attempting when he said "haste makes waste" or "a penny saved is a penny earned." It was the same problem that his elders were addressing when they unsuccessfully tried to persuade my grandfather to remain Amish, and stick with the buggy and plow with horses. As it turns out, our Christian ancestors didn't make progress here on Earth as much as they did a mess. Certainly few of our lives materially and spiritually have to do with the teachings of a man called Jesus. I went to school and there I remain, although I have learned more about respect for all life from people who have spent less time in our modern classrooms and are thus supposedly less educated.

Conclusion

We can never hide from our responsibility for being a part of everything happening in the world in which we live. Accepting that is at least a precursor to experiencing greater understanding and spiritual enlightenment. As Cat Stevens' song suggested to us in the '60s, "we are on a road to nowhere, we best should take our time." When your car's gas gauge is near empty, the worst thing one can do is to speed up trying to quickly get to a gas station. Thus when we see people slowing down trying to conserve non-renewable resources, we should support their efforts, since less energy spent means there will be more available for all the life that is here now and all the life that comes hereafter.

The sights and sounds I experienced on the road to Cuba reinforce my faith in the ultimate moral rule, which is to use as few resources as possible and in so doing express our love for one another, our love of life and our commitment to its continually-unfolding state. Some say the highest form of love is self-sacrifice, the willingness to go without, even to give one's own life, if necessary, to foster life itself. When I was small, I heard stories of Anabaptist heroes who held such views. But as I grew, I became educated into another set of values, and like many of my sect, I desired above all to be considered a success and thus sold out to a cultural system of ethics that believes more and bigger are better and often stands ready to kill to get a raise.

A New Beginning

On my trips in Latin America I often stop at Catholic churches. They always seem to be open. They are cool and quiet and a good place to go to meditate. Sometimes there are pigeons and other birds inside and always there is a figure of Jesus nailed to a cross. Both the pigeons and the image on the cross serve as symbols for me. The pigeons symbolize the life we still have. The figure on the cross symbolizes that this life will physically but not spiritually end, and *how we live here and now*, no matter where on the road we think we are at the moment, or where we think the road may lead, is what judgment is all about.

Footnotes

Boyd, D. (1974) *Rolling Thunder,* Dell Publishing Company, Inc., New York, New York.

Castro, F. (1985) "This is the Battle for Latin America's Real Independence," *Address given in the 4th Congress of the Latin American Federation of Journalists (FELAP) during the afternoon session of Saturday, July 6.*

Castro, F. (1985) "The debt is not only unpayable but also uncollectible," *Address in the 4th Congress of the Latin American Federation of Journalists (FELAP) in the afternoon of Sunday, July 7.*

Dunn, William (1986) "Commune Adapts to the 80s," *The Detroit News,* April 13.

Gross, B. (1980) *Friendly Fascism: The New Face of Power in America,* South End Press, Boston.

Haenicke, Diether, H. (1988) "Education and industry can join to stem slide into illiteracy," *The Kalamazoo Gazette,* May.

Rifkin, J. (1981) *Entropy,* A Bantam Book.

Skinner, B.F. (1948) *Walden Two,* Macmillan.

Ulrich, R., Skinner, B.F. on The Dick Cavett Show (1969), New York, New York.

We Are All Related to One Another

Reprinted from The Kalamazoo Gazette, *November 12, 2008*

A friend of mine recently spoke of his angst relating to the influx of folks coming across United States borders. This concern was shared by both the Republican and Democratic presidential candidates as they discussed what should be done about "illegal" immigration, as well as how long our troops should stay after our own "border crossings" into Afghanistan, Iraq and various other countries around the world.

Whether we like it or not, we seem to be on the wheel of "what goes around comes around," wherein all of us, like spokes, are attached to both the rim and the hub, where we unite as one. All people on Earth are related. We are all citizens of Earth. My genetic family came to North America on the back of laws made by people the natives saw as invaders.

In 1630, John Winthrop led a group of what the natives looked upon as immigrant aliens onto the shores of what is now called North America. At that time there were about 700 in all who, as Puritan Christians, constituted the largest party of immigrants to have made it to the so-called New World. More were soon to follow. In short, a growing group of deeply-religious folks from Europe stood ready to build what they felt would become a "city on a hill," ruled by the principal idea that their "supreme deity" loved each person equally.

Hidden, however, within the rhetoric of that belief was the existing hypocrisy that there was no intention of having the Native Americans, nor for that matter the women in their group, play an equal role in deciding what was to transpire.

Today, as we watch Republicans and Democrats debate the matter of the immigration of undocumented aliens into the United States, we might again consider that it is certainly not a novel issue. Indeed, many Native Americans who were here long before my great-grandparents came to occupy the natives' cultural and spiritual home, no doubt wish in hindsight they would have handled differently the question of who and who not to allow into "their" country.

As is being shown in Afghanistan, Iraq and other places around the globe, it actually matters little who was here or there first. Who is legal and who is illegal depends more upon who supposedly won the latest battle. People move around from country to so-called country, and what constitutes "legal" depends upon who makes up the current laws and who has the resources available to hire the judicial system and the guns that back the current powerbrokers' opinions of what constitutes equal justice.

As the political debates between Republicans and Democrats go on, it would be wise for us to remember that there is a bigger body of law not of man's making and not impressed by human arguments or campaign promises.

These are the laws of nature and the sometimes inconvenient truths that will, in the long run, determine all our fates in the final court of natural justice. These are the laws under which we ultimately live as we come face to face with the here and now way that things are. At each day's end perhaps the least we can do is pray for help from the great gift-giver to assist us during the years ahead to live together impeccably and with as much integrity as we can muster as we approach the border over which we all finally cross as members of the same party and travel into the next manifestation of this strange mysterious here and now we all share.

What Goes Around Comes Around

Letter to the editor of The Kalamazoo Gazette

Dear Editor,
    The Gazette's *editors recently printed a viewpoint by the* Cleveland Plain Dealer *entitled: "Homeland Security owes explanation for security breach." In my opinion, it is perhaps the case that the breach, which was discovered aboard a jet on its way to Detroit in 2009, can be explained by the events surrounding the 1492 landing by Columbus in the Bahamas. My Native American friends often express regrets over the laxity of the security and the naive immigration laws that were in effect at that time among the Indians.*
    What Columbus did to the Arawaks of the Bahamas, Cortes did to the Aztecs of Mexico, Pizarro to the Incas of Peru and the English settlers of Virginia and Massachusetts to the Powhatans and Pequots. The terror, genocide and broken treaties which spread across the Americas following the entry of what proved to be dangerous aliens to this land is in many ways mirrored by current events, which might also be explained by the cliché "what goes around comes around."
From,
Roger Ulrich

The Use of Behavior Modification Strategies to Increase the Probability of Attendance at Evening Chapel Through the Use of Food Contingent Reinforcement at the Life Line Mission, San Francisco, California

Reprinted from Behaviorists for Social Action Journal, *Vol. 2, No. 2, 1980, pp. xxix-xxxiv*

He asked me for 25¢ for a cup of coffee in what turned out to be a Pittsburgh accent. I said, "Sure, if I can have one with you." I had just hitchhiked into San Francisco. He had been there for a month. I had a room that cost me $22 for the week; he used the parks, the alleys, under the freeways, and now and then the missions. His eyes twinkled and he smiled a lot. He said he was afraid of the country, afraid of snakes and that he had always lived in cities. I said that I wasn't afraid of snakes as much as I would be of sleeping under a freeway or in an alley in San Francisco. He laughed.

I finished my coffee and he his, and I said, "Maybe I'll see you again sometime brother, so long."

I got on the cable car at the foot of Powell St. and rode over the hills to where Alcatraz can be seen and the tourists come to watch the natives dance, sing and sell their wares. The sun shown down all day as I sat and reminisced, while a group of three played and sang Jackson Browne songs midst the chaos of banjos, trumpets, Hare Krishna bliss, and panhandlers.

At about 5:00, I started walking back to my hotel, the Herbert, stopping in Chinatown for some shrimp chow mein and a few more moments in the sun in Washington Square and a few more moments inside the Catholic church, there just being still while a handful of the faithful came in to pray.

It was late when I got back and went to bed. Monday was breaking when I got up for the third time, this one to stay. I exercised, meditated again and called an old friend. We met on the 34th floor of the Bank of America Building, which was completely held down by a law firm which is rich because they do a good job defending insurance companies against those who make the claims.

My friend and I had a nice lunch. We talked of the old times, the drug bust, the matings and the re-matings—the hard feelings that were left and the love that was left and how good it was to see each other again. Then an afternoon at the Embarcadero with the sun and the street sellers.

After a quick wash-up in my room plus some minor chores, I found myself walking around Union Square.

"Hi Roger."

"Hi Richard," I said as he twinkled. We shook hands like old friends.

"Have you been to St. Ambrose yet?" he asked.

"I'm going down to the Life Line Mission on Fulton and 5th at 6:30. You want to go? It's a free meal. All you have to do is listen to an hour sermon."

"OK," I said.

The front of the mission was darkened by about 40 men, with more filing in from all directions. It wasn't quite like a prison scene or a mental hospital or the Navy. The men were less uniform, not as clean, not as orderly. I had the feeling they were more free. An old man of who-knows-what age slept leaning against the side of the building. The clothes on many of the people were all they had, and they wore them night and day. A few had small suitcases, a bed roll here and there, a couple of backpacks. It was pretty quiet. The doors opened and everyone pushed quickly to the inside. The 60 seats were soon filled.

"Can I take a leak somewhere?" I asked.

Richard said, "Can't you hold it?"

I said, "No," and walked over to a young man guarding a door.

He said, "OK, if you got your seat marked."

"It's marked," I replied. "Where is the head?"

When I returned, a man of 30-35 years, with a mustache, a clean shirt and a grim look was calling out rows and handing out bed assignments for the 14 bunks that were available that night.

Things moved slowly but finally the service began with an introduction of the minister. Some of the men were asleep, many were semi-drunk, most were totally inattentive. In short, they appeared to be simply waiting for the hour to be over so they could eat. After all, they had heard the same "you're-a-sinner-come-to-Christ" speech for years now as they did their hour of Mission time response for the meal reinforcement.

"How many of you feel like singing tonight?" asked the preacher.

A couple hands went up and some groans. The singing began. A few really got into it, while most of the rest continued to patiently play their waiting role. A young man from some local youth group got up and told how he had been saved in a nervous voice that made him seem unsure of whether he still was. Finally he sat down.

His audience seemed, through every sense I possessed, to be the most seasoned group of humanity I had met for sometime. They listened politely. The urine smell mixed heavily with the odor of too much alcohol and too many days without washing.

Then came the preaching, which got heavier and heavier as the hour wore on until the mumble of "Oh shit, shut off" from behind me and to my right began to come more often.

"You must all realize that the Devil has hold of you, and through Christ alone can you pull yourselves from the mud and the slime."

"Oh shit man, wind it down," came the reply, a bit louder now as more of the group began to grumble slightly.

The young manager who had handed out the beds began to look nervous when the minister announced that he had "five more minutes and was going to use them because they might be five minutes that would finally bring some sinner to his senses, senses dulled by alcohol and lives without Christ."

I began to feel uncomfortable and embarrassed—not for the men among whom I sat as we waited for our hot soup, lemonade, donuts, bland lettuce, and hash—but for the man up front who looked down on us as those wretched sinners out there. The prayers were for us the sinners. The sermon was for us the sinners. The songs were for us the sinners.

"Do you like it in the gutter?" the preacher asked.

"It's as good a place as any when you're high, man!" said one.

"Hey, let's eat!" replied another.

"Well, that's where you will stay. The Devil has got you," announced the preacher.

"Oh shit!" said a voice from behind me and to my right.

Has he, I thought. Whom has he got? That morning I had lunch in a smart spot by the Bank of America Building among the beautiful people whose crimes are well accepted, as are always the crimes of being rich. The old Chinese man beside me, just out of prison, was an alcoholic too poor and not educated as to how to get his drugs and remain respectable. For me, it was like being a child again—having to sit quietly through a Mennonite service because, if you didn't, your reward might not come in heaven and for sure it wouldn't on the way home from church!

In the front of the Mission chapel there was a painting done perhaps by the top artist of the First Baptist Church of San Francisco. It was a scene depicting two roads, the wide way which led off toward a big fire and a narrow way that led off and over a mountain. The roads were marked "wide way" and "narrow way." The verse below is common to all who have heard the Christian message: "Many will go the wide road, only a few will follow the narrow."

Richard and I left the mission, where we 60 sinners had sat and had been preached to as we waited for our meal, and began to walk the late evening San Francisco streets. And I thought and I wondered, as I watched growing numbers of night citizens parade in their fine clothes, fine cars going to and from expensive homes between expensive ships, buying gifts, going to shows and to drink at the Top of the Mark, just who is on the wide road going toward the big fire?

Richard said, "Good-night, Roger," and twinkled as he went off in the only set of clothes he had to sleep under a U.S. 80 overpass.

I said, "Good-night, Richard," and went back to my hotel to sleep the night. The American Psychological Association convened the next day in the expensive hotels where I was to introduce a colleague. I thought about how we, as psychologists, would be talking about solving behavior problems and found myself wondering if perhaps we were as much the problem as we were the solvers. I thought about sitting in the Life Line Mission with Richard, being preached down at in order to get a reinforcer, and of the basic strategy of behavior modification in which we frequently assume that we (as behavioral technicians, clinical psychologists, therapists, or whatever other important-sounding title we happen to be using at the time) know what is "good" behavior and withhold reinforcement until it occurs in those less-informed people we are attempting to control. At the convention there would be many talks on how we professionals could solve big world problems. While pondering all this I recalled something Einstein once said, "The world we have made as a result of the level of thinking we have done thus far creates problems that we can't solve at the same level as the level we created them at." As I began to doze off, the sense of the Einstein quote was beginning to break through—the only way our human problems can be solved is by creating a new level of thinking about them. How that new level was to be created, however, wasn't totally clear. Then I remembered having heard that in the United States, 5% of the world's population uses about 45% of the world's energy. Many of the psychologists at the convention work at solving the problems of the poor. It may be that Richard is closer to the solution than we psychologists. My room in San Francisco cost $22 a week. At the Hyatt Hotel off Union Square, where many of the psychologists stayed, some rooms went for $150.00 a night. Maybe I'll meet another Richard soon, I thought as I finally fell asleep.

On the way home from San Francisco I met and stayed a week with an Indian medicine man named Rolling Thunder. He said, "You have to live the truth and be a part of it and you might get to know it. I say you might—and it's a slow and gradual process and it don't come easy."

I often wonder where Richard slept last night.

Notes from a Radical Behaviorist
A Brief Case Study on a Number of Persistently Disturbing Sexual Irritations and Some Questions Pertaining to the Drift Toward the Formation of Another University Committee to Revise the Ten Commandments

Reprinted from Notes from a Radical Behaviorist, *Vol. VI, Issue 2, 1986-87*

Dr. Haenicke's address on "Ethics in Academe" and Dr. Malott's essay "In Defense of Sexual Harassment" have prompted some thoughts which I wish to share with the readers of "Notes" in the spirit of academic freedom and truth. Dr. Malott indicates that his dictionary says that *"harassment"* is a persistent disturbance or irritation. So I looked in mine and, between harlotry and hast, *"harrassment"* was missing. (*Webster's Seventh New Collegiate Dictionary,* 1972). From harquelius (an obsolete portable firearm), Webster's crew skipped right past "harrassment" to harridan (a scolding old woman). So in a scholarly fashion I went to Webster's (1966) third *New International* (which is hard to lift), and there between harquelius and hash, I found the truth. I was misspelling harassment (with two r's) and thus returned to *Webster's Seventh New Collegiate,* 1977, and there between harangue and harlot was harass. vt. [F harasser, fr. MF, fr. *harer* to set a dog on, fr. Of hare, intery used to incite dogs, of GMC origin; akin to OHG heir here - more at Here] 1: to worry and impede by repeated raids   2a: Exhaust, fatigue   b: to annoy continually, syn see worry.

Well that was just a small dictionary and I was already getting confused. It didn't at all say what Dick's dictionary said. So I went again to the heavy Webster's Third, and there on page 1031 between happen and hard, was haras 1 after archaic: a horse-breeding establishment; stud farm 2: Harras (like I started to spell it in the first place). Well they made it into a long story. It talked about laying waste (as an enemy's country; Raid HARRY < hostile Indians, the frontier. It brought up worry again, and talked about guerrilla forces who cooperated with the United States parachute troops, e.g., the Japanese. It mentioned exhaust, fatigue—William Wordsworth to vex, trouble, or annoy continually or chronically as with anxieties, burdens, or misfortunes, plague, bedevil, badger.... It even mentioned a *lack of funds*, and hell I was still a full inch and one half away from harassment!!

Well, since the dictionary was written in 1966, and given the preposterous detail, I figured that maybe it was all done by a bunch of hippie grad students on LSD anyway, so I decided to put it back and returned to my scholarly pursuit. Dick's definition sounded okay by now, and certainly the frequent linking of "sex" and "harassment" seemed to deserve further exploration and of course further discussion!

As best as I can remember, I experienced my first persistent, sexually-disturbing irritation when I was about five. I was staying with my Amish-Mennonite grandparents and I was assigned at bedtime to sleep with one of my good-looking aunts. I recall that before going to bed, I had been looking at pictures in a National Geographic and fantasizing my role as the savior of the beautiful Aztec maiden about to be tossed into a flaming volcano. I spent the night in a cuddly mood. The next morning, my aunt allowed in a slightly distressed voice:

"You were pretty 'lovey dovey' last night, weren't you?"

I remember feeling embarrassed and guilty because she said it with the same kind of tone and expression that she and other relatives used when I had done something to disturb them. (Whether or not I was her "harasser" or the editors of the *National Geographic* mine, I will leave to some university committee to decide, or maybe a dean, and agree or appeal depending upon their decision.)

I recall having felt a similar irritation some years later when a bunch of fourth grade boys, at the Eureka Davenport Grade School, held me down as a first grader so that Bertie Lou Highway could kiss me. I was so in love with Bertie I couldn't see straight to read about Dick and Jane and Spot, but couldn't admit it and so I felt mixed emotions about the "harassment."

Ann Lathrop brought me a May basket in the fourth grade and I ran after her as per custom, which decreed that if I caught her, I had to give *her* a kiss. She ran as she was supposed to, but then suddenly she stopped, turned toward me grinning, beckoned persistently, and when I came to her, she gave *me* a big kiss. Afterwards, I got so silly that I rode my new Schwinn down Hamilton Street standing on the seat, with no hands until I hit some loose gravel and had a very disastrous wreck that left me with some pain that persisted for about a week. That was at least two years after Jack Stromberger and I had had our long discussion on the way home from school and had come to the conclusion that our folks probably did it just like horses, cows and dogs, which left me a bit disturbed since it seemed preposterous and more than a little bit dirty.

Then there was the Sunday at the Roanoke Mennonite Church when someone read the Bible verse in Viola Zook's class about "if your right hand or eye offended you, you were to cut it off or pluck it out," since it would be better to enter heaven whole than go masturbating and ogling past the gates of hell! I really tried hard to stop harassing and abusing myself for a while after that until there was sort of an accident one night in a parked car with a second cousin from Goodfield. It involved an intense, persistent irritation that resulted in such an embarrassing disturbance that my mother accused me of sexual misconduct when she observed the condition of my Sunday suit pants, which I inadequately tried to hide on the way to the cleaners.

In 1950, after my first year of college, I went to Europe for the summer with a church group and saw the Pope, a lot of bombed-out buildings, and met a bunch of German kids I liked. I was 18. Also, there were a lot of female teachers in our group between the ages of 24 and 62. It was a fantastic three months, with a lot of sexual conduct, some of which no doubt bordered on being prefixed with *"mis-,"* depending on who was doing the labeling. Anyway, let me try to make what could become a very long story short.

Somehow, something or other in my background kept me from the ultimate sin of the unmarried until I got out of the Navy. Amish-Mennonites were not supposed to fight either…but since I had been in so many during my early life, I didn't think that either the Selective Service board of Woodford county or the F.B.I. agents (who were always snooping around my home town flashing their badges and asking all kinds of questions about my male relatives and friends who had guts enough to refuse to join the armed services) would believe me if I tried to convince them that in addition to not having sex…I wouldn't fight. So I enlisted. Two years later, after flunking out of Officer Candidacy school, and after the end of the Korean War (which was the reason for my going off to learn how to kill people in the first place), I came home. Now with two years of experience floating around oceans as tanker deck-ape, I set forth to get my masters degree…my first real job and my first taste of breaking the *8th* Commandment.

She was beautiful, 21, respected, Methodist and a student when we somehow came to terms with what was a mutual agreement to sexually harass each other. I was 24, a newly hired assistant dean of students and assistant basketball and baseball coach at a respectable ivy-covered college. She sang opera, ran for homecoming queen and was a member of the university judiciary committee. She was even more beautiful and respected when we were married three years later and were told in front of all our relatives that this was going to be for better and/or for worse. She is even more beautiful and respected today, some thirty years and three adult children later, after other seductions and a lot of harassment (both vice and versa) plus, for sure, a lot better acquainted with what is meant by *"for better and/or worse."*

True sexual harassment is an abhorrent aberration no matter where or when it occurs. The human species has raped, exploited and harassed Mother Earth in ways that no other life form has remotely approximated via countless actions, many of which were first thought of at universities. We, as a species, have evolved to a place with such little respect for Mother Earth that we often find some of our members treating each other with similar disdain. If somehow another university committee needs to be put together to reword the Ten Commandments, well that's just life in academe, and of course there is always the possibility of a federal grant to fund it. Perhaps if we somehow tag it on to a Star Wars proposal, or make it pertain especially to R.O.T.C. training or recruitment (like in those TV spots which entice kids into the armed services so that they can be trained how to go kill other kids), it will get special attention. Personally, I think the Ten Commandments are all we need and besides, when you look closely, people don't pay any attention to them much anyway.

However, I am, to a certain extent, a realist and we do need to keep busy at universities researching, publishing, etc., or else maybe lose our paychecks, and so if a joint administrative-faculty committee is deemed necessary to pursue further issues which are felt not to be adequately covered within the Commandments, then I say let's go for it and nominate Deither and Dick as co-chairmen.

CREATIVE EXPRESSIONS

The Process Called Life

Reprinted from Touchstones on the Path: Some Experiments in Living, *Life Giving Enterprises, Inc., Kalamazoo, Michigan, 1988*

I am conceived,
                I grow,
                              And I die.

And in those moments
Between the conceptions and the deaths,
There occurs the change
Of events which, in a sensible way, take the stuff
Of eternity's evolution
And mold it into the actions
That I learn to know as me.

Though still
                I understand not,
                              Neither why nor how...

Genesis Revisited

Reprinted from Touchstones on the Path: Some Experiments in Living, *Life Giving Enterprises, Kalamazoo, Michigan, 1988*

And people talk about the beginning and I wonder,
beginning of what? I think there was no beginning. No
creation. No "Let there be this," no "Let there be that."
No Beginning!
        Only the forever
                And the forever
                        Of eternal change.

Rites of Passage

For children born into the Anabaptist Amish, Mennonite culture, the most important rite of passage was that of "joining the church." When I was a child, the first step in the ritual was often performed at the end of a revival meeting, when the call was given to "Come to Jesus." So when the invitation was given for all sinners wanting passage on the road to redemption to rise, I stood. Dad quickly put his hand on my head and pushed me back down to sit it out for a bit longer. Age five was too young for deliverance. I guess my Dad knew I had more sinning to get under my belt. At age 10 I rose again when promised atonement by Reverend Erb for admitting my transgressions. My decision to stand and confess was framed by the Amish Mennonite congregation soothingly humming "Oh Lamb of God Descend On Me," fear that I would go to Hell if killed on the way home from the revival meeting, and the peer-pressure from my cousin Harold, who I was sitting with, having already popped up! Given what was taught in Sunday school about what kinds of behaviors would get you into Hell, a future for me there loomed as a distinct possibility. When a young aunt came running up after the service laughingly asking, "What got into you?" I took it to be a sign that there might be more to this game than what meets the eye.

In 1965 I was invited to Harvard as a Ph.D. research scientist speaking on strategies of behavior control. One colleague showed how an angry bull could be calmed down when given a slight shock to its posterior hypothalamus. I reported on how punishing aggression produced only more of the same, and another researcher discussed the effects LSD had on schizophrenics. When I asked him what taking LSD was like, he said he hadn't tried it, but other Harvard professors had. When I returned home, I asked a student to get me some, which when ingested provided me with yet another rite of passage!

Some years later, shortly before Christmas 1971, I mainlined cocaine and simultaneously swallowed some blotter acid. While coming down from the cocaine, the LSD kicked in and I departed on a trip that included all creation passing before my eyes. As I zoomed by what my genetic and environmental past had left as residual, my mind was racing so fast it simply broke, and everything stopped. Amidst agonizing terror, I realized I was in Hell and unable to talk my way out. Things then got worse. My arms, where I had injected the cocaine, throbbed, my stomach cramped with pain that increased moment to moment, and it all continued for eternity! Then came the realization that I'd died, and a feeling of peace came over me as I nestled into a state of calm, which shifted every so often to still another sense of "this is how it will be now and forever." Later, I realized I wasn't dead, but instead simply another "born-again" child of life still on the road toward the next bend.

I'm now 70 and more cautious about what I do with my body. I serve my current atonement trying to live well with less on a semi-communal homestead that looks like something out of the third world trying to survive a continual depression. My body and the land upon which I live is my church, my life is my religion, and I follow a spirit quest wherein I've grown to appreciate the concept implied in the song "Amazing Grace," which symbolizes that it is not totally within our control to determine when one will be allowed the privilege of facing still another rite of passage.

A Gift from Gawath-ee-lass

She wore feathers in her hair and danced in the rain on the way toward the school that tried to break her spirit...and no one seemed to understand this alien in her own land... A hidden seed of Natives who refused to die.

Now disguised in foreign body and dress, she grew and learned the customs of the civilization that killed her fathers and her mothers.

Lie low, whimpered the raccoon. Don't cry, said the wolf, hidden in the body of a fox terrier. We must wait patiently in our prisons of flesh and mind for the keys to be turned so that our spirit may be freed to float again like the bread of life over the land of our forebearers, poisoned by the greed of those who don't understand the need to respect all our relations.

Struggling in her cocoon, she hears in the distance the morning prayers to Father Sun and Mother Earth, spoken in the wind, in the rivers and in the tongue of her kin on Great Turtle Island, who like Peltier stand imprisoned in the concentration camps scattered about a land called free.

Your time will come, butterfly. As prison executives grow rich from the business of imprisonment, the windows of truth will one day open again and the keepers of the Earth will invite everyone's presence into the question of balance. How and where you will lead was settled long ago...the Earth needs your love, the White Buffalo has been reborn.

Drumbeats call forth the days of purification in the hearts of those for whom the Great Mystery has assigned the medicine role. There is no escaping the gift.

Reflections from the Martyr's Mirror

Part 1

Reflections from the Martyr's Mirror
   Showed how strong hearts break
      As plows killed off the Indian
         And behind a mask of fake
            Aliens crowned "God's chosen"
               For their own selfish sake
                  Penned bills of inequality
                     That legalized their take

Then on a boat from another land came an Anabaptist child
Trained to be a Martyr, a victim, meek and mild
Led by the Ten Commandments to forget he once was wild
Who when his church demanded that their rules he must obey
Swapped his buggy for a car and drove another way

Now in the world of having fun
   With other rogues on the run

        He learned to shoot an M1 gun
            In the name of the god who killed his son
                Then marching left and sometimes right
                    Like a patriotic dog
                        He went to wage the war
                            That spreads hate out like fog

Now everywhere that judges go
    Their judgments follow me
        A hunter-gatherer "turn-coat"
            'Neath the stuff that's PhD
                Which choked the life from the land
                    As banks forced us to flee
                        Into the arms of modern ways
                            And the myth that we'd be free

Then one day I heard the hymn that prisoners sing off-key
Echoed from the wild wolf's howl which made it clear to me
That I'd been tamed just like the dog the wolf turned out to be

            So the moment came that all men fear
                Midst terror that peace forgot
        When news spread out from Nickel Mines
                Where Amish kids got shot
        And there instead of hope being lost
                Forgiveness gave and it was not

So back I went to the future
Away from city lights
To commune again midst songs of frogs
And Mother Nature's sights

Now in a "spirit being"
Disguised as human hide
I ran back to my buggy
For the homecoming ride

    Where I watched myself reflected
    In the shadow of the sun
    That follow the trails of how
    On earth your life is done

        Where alone we walk together
        As pilgrims come and go
        Waiting for the curtain drop
        To get set for another show

So when you stand in judgment
    If the life you lived was fair
You're handed "grace amazing"
    From the Mystery we all share
In the everlasting present
    Labeled "handle with great care"

Then in the mirror eternal that reflection comes again
Of the self you're always meeting on the road that has no end.

The River of Relief

Prologue

One by one the stings hit the wolf as he returned with great purpose to the river of relief whereupon he paused after caressing his wounds and gave thanks to the four directions and the Father Sun and the Mother Earth before eating the last of the several hornets that had caught themselves in his thick coat and he recalled again the way life works…take that sting…and that kiss too…and take that kiss…and that sting too…

Scene I

At that very moment seemingly at another time and place the masters of take sat talking around the table of greed and exploitation attempting to be seen as wise leaders important to whatever the current theft at hand extorting pretenses of glory to broken promises as their armies marched killing as they went while the mother of the only son sat waiting by the dying coals of the human fire which announced one by one the names of hunters gatherers and other species long gone from where truth once flowed wild from lost languages that no one now understood and once hallowed sights that no one now dares see as the developers sprawl cut deep into the

The River of Relief 313

earth's side like a mean trick from the grand illusion of being a part of the tribe yet unworthy of the holy father's kiss for having hauled the wood to the place of the secret meeting where she was forced to spread her legs for the explosive sticks that spit an alien flame into the birth womb that brought forth the first rogue primate and her sadness for all that it meant for generations to come....

Scene II

Dare I cry out the mother cried out and why am I here on my knees being judged by all my relations who helped force me to labor all the days of my life beside the tables of take and rape for all the things they wanted their way for all the things that in time gives way and please forgive me if I am speaking too loud she said to the wolf as he emerged from the river's edge I fear that my shouts for living the truth of life's golden rule that we honor the air the water and the earth which does unto us as we do unto it will cause problems for my people and for you four legged ones and I fear my kind will call me a pagan of no value soon to be forgotten like the last stone dropped on a lonely road leading nowhere please help me wolf as I walk past the shadow of my death and I pray to fear no evil in the name of the father son and the holy ghost....

Scene III

Good god said the wolf as he shook off the river's water what a mean trip civilization has run down on you and may the great spirit help you to not fear for you're a fine being who was once like me doing no more or no less than what was done to you and please do not allow any more trespass against your wild heart and do not let anyone domesticate you any further I know it will be hard for when it comes right down to the truth of it that is why I too feel so insecure and confused when I see what happened to the poodles the rat terriers and the other dogs who were once wolves like me and the rogue primates who got tamed and civilized is it any wonder that I howl at the moon at night and in the morning as the sun comes up and then again at noon so lonely and so fearful that peace leaves my heart and I howl to nature to help me and it seems I just can't stop you know how it is you're a mother who cries and you have your only son whom you hear cry and I too cry like you and like him when I'm desperate scared and terrorized as they shoot at me and set out traps and try to kill me to use my fur to make the coats and blankets that are bought and sold on the world market and when they put me in zoos to show off the conquest of the wild as I pace back and forth like the sad bear and the lonely gorilla and oh how it hurts when they tell stories to children to be afraid of the big bad wolf and when ladies like you are told to watch out for the wolf in sheep's clothing who wolfs down food that they refuse to eat and I feel so sad when loved ones are told not to associate with the likes of me if they want to keep their politically correct reputation of virtue… and the wolf wept…as they say did Jesus…..

Scene IV

Finally the wolf paused breathed deep and smiled and then she smiled and their eyes shined as together they walked over to a log where they sat and watched as the wild geese flew into the evening and they remembered again that time would pass and differences would come and go and one day a voice came in a whisper which pulled the trigger on the memory of all the things the wolf had told her how its scary when you're shot trapped killed skinned and stories told to children to be afraid of the wild wolf in sheep's clothing that runs with the coyote who illegally sneaked across the borders of time past…..

Scene V

Before long the sun began to set and the whispers grew louder until finally they turned into shouts and her ears rang with terrorizing fears and doubts and questions were raised about associating with the likes of such an animal if she wanted to keep her reputation and she was told to not hang out with the likes of the wolf and other illegal aliens for their days were numbered and that they were like dirty dogs and she was told to enroll in the schools that teach how to keep the wolf away from the meetings at the tables of modern technology and how to keep him in the fields growing the cheap food which tastes so good and which is made so easy to get and she was told that she was well suited for modern life and its comforts and she was told that the wolf was soon to be eliminated and that the assassins were already poised to kill him and not to listen to his howling yips and whines as to how the disasters from which the civilized primates claimed to save the world were not as the wolf had claimed as problems that came with the rogue primates and which followed them from the day of their conception and she was told she needed to consider her reputation for there were already those out on the street saying that she was more than just the wolf's friend....

Scene VI

Okay okay okay she cried we need to talk wolf I've heard your story over and over time after time you are like a broken record howl howl howl yip yip yip whine whine whine I who am strong and vulnerable gentle and fierce I who see with the clear vision of one who knows we belong only to ourselves I need a break from you for the moment has come for moving on and as a special favor for our years of friendship and caring please stay away from those who speak at the corporate earth day parties for they tell me they are bothered by the sight and the sound of your wild howl and your companions who like you stoop in the fields and sneak back and forth across and under the borders so I must go now and walk the civilized path that calls me away from the spirit of the free and unspoiled wildness we once shared I will miss you as you miss me until we meet in that mirror of time where reflections come again of the self we're always finding on the road that has no end.....

Epilogue

One by one the stings hit the wolf as he returned with great purpose to the river of relief whereupon he paused after caressing his wounds and gave thanks to the four directions and the Father Sun and the Mother Earth before eating the last of the several hornets that had caught themselves in his thick coat and he recalled again the way life works…take that sting…and that kiss too…and take that kiss…and that sting too…